Muds and Mudstones: Physical and Fluid-Flow Properties

Geological Society Special Publications
Series Editors
A. J. FLEET
R. E. HOLDSWORTH
A. C. MORTON
M. S. STOKER

It is recommended that reference to all or part of this book should be made in one of the following ways.

APLIN, A. C., FLEET, A. J. & MACQUAKER, J. H. S. (eds) 1999. *Muds and Mudstones: Physical and Fluid Flow Properties*. Geological Society, London, Special Publications, **158**.

DEWHURST, D. N., APLIN, A. C. & YANG. Y. 1999. Permeability and fluid flow in natural mudstones. *In:* APLIN, A. C., FLEET, A. J. & MACQUAKER, J. H. S. (eds) *Muds and Mudstones: Physical and Fluid Flow Properties*. Geological Society, London, Special Publications, **158**, 23–43.

GEOLOGICAL SOCIETY SPECIAL PUBLICATION NO. 158

Muds and Mudstones: Physical and Fluid-Flow Properties

EDITED BY

ANDREW C. APLIN
(University of Newcastle, UK)

ANDREW J. FLEET
(Natural History Museum, London)

and

JOE H. S. MACQUAKER
(University of Manchester, UK)

1999
Published by
The Geological Society
London

THE GEOLOGICAL SOCIETY

The Geological Society of London was founded in 1807 and is the oldest geological society in the world. It received its Royal Charter in 1825 for the purpose of 'investigating the mineral structure of the Earth' and is now Britain's national society for geology.

Both a learned society and a professional body, the Geological Society is recognized by the Department of Trade and Industry (DTI) as the chartering authority for geoscience, able to award Chartered Geologist status upon appropriately qualified Fellows. The Society has a membership of 8600, of whom about 1500 live outside the UK.

Fellowship of the Society is open to persons holding a recognized honours degree in geology or a cognate subject and who have at least two years' relevant postgraduate experience, or not less than six years' relevant experience in geology or a cognate subject. A Fellow with a minimum of five years' relevant postgraduate experience in the practice of geology may apply for chartered status. Successful applicants are entitled to use the designatory postnominal CGeol (Chartered Geologist). Fellows of the Society may use the letters FGS. Other grades of membership are available to members not yet qualifying for Fellowship.

The Society has its own Publishing House based in Bath, UK. It produces the Society's international journals, books and maps, and is the European distributor for publications of the American Association of Petroleum Geologists (AAPG), the Society for Sedimentary Geology (SEPM) and the Geological Society of America (GSA). Members of the Society can buy books at considerable discounts. The Publishing House has an online bookshop (http://bookshop.geolsoc.org.uk)

Further information on Society membership may be obtained from the Membership Services Manager, The Geological Society, Burlington House, Piccadilly, London W1V 0JU (Email: enquiries@geolsoc.org.uk; tel: +44 (0)171 434 9944).

The Society's Web Site can be found at http://www.geolsoc.org.uk/.The Society is a Registered Charity, number 210161.

Published by The Geological Society from:
The Geological Society Publishing House
Unit 7, Brassmill Enterprise Centre
Brassmill Lane
Bath BA1 3JN, UK
(*Orders*: Tel. +44 (0)1225 445046
Fax +44 (0)1225 442836)
Online bookshop: http://bookshop.geolsoc.org.uk

First published 1999

The publishers make no representation, express or implied, with regard to the accuracy of the information contained in this book and cannot accept any legal responsibility for any errors or omissions that may be made.

© The Geological Society of London 1999. All rights reserved. No reproduction, copy or transmission of this publication may be made without written permission. No paragraph of this publication may be reproduced, copied or transmitted save with the provisions of the Copyright Licensing Agency, 90 Tottenham Court Road, London W1P 9HE. Users registered with the Copyright Clearance Center, 27 Congress Street, Salem, MA 01970, USA: the item-fee code for this publication is 0305-8719/99/$15.00.

British Library Cataloguing in Publication Data

A catalogue record for this book is available from the British Library.

ISBN 1-86239-044-4
ISSN 0305-8719

Typeset by WKS, Westonzoyland, UK

Printed by Alden Press, Oxford, UK

Distributors

USA
AAPG Bookstore
PO Box 979
Tulsa
OK 74101-0979
USA
(*Orders*: Tel. +1 918 584-2555
Fax +1 918 560-2652
Email bookstore@aapg.org)

Australia
Australian Mineral Foundation Bookshop
63 Conyngham Street
Glenside
South Australia 5065
Australia
(*Orders*: Tel. +61 88 379-0444
Fax +61 88 379-4634
Email bookshop@amf.com.au)

India
Affiliated East-West Press PVT Ltd
G-1/16 Ansari Road, Daryaganj
New Delhi 110 002
India
(*Orders:* Tel. +91 11 327-9113
Fax +91 11 326-0538)

Japan
Kanda Book Trading Co.
Cityhouse Tama 204
Tsurumaki 1-3-10
Tama-shi
Tokyo 206-0034
Japan
(*Orders*: Tel. +81 (0)423 57-7650
Fax +81 (0)423 57-7651)

Contents

APLIN, A. C., FLEET, A. J. & MACQUAKER, J. H. S. Muds and mudstones: physical and fluid-flow properties — 1

Physical properties

PEARSON, F. J. What is the porosity of a mudrock? — 9

DEWHURST, D. N., YANG, Y. & APLIN, A. C. Permeability and fluid flow in natural mudstones — 23

MIDTTØMME, K. & ROALDSET, E. Thermal conductivity of sedimentary rocks: uncertainties in measurement and modelling — 45

PETLEY, D. N. Failure envelopes of mudrocks at high confining pressures — 61

BJØRLYKKE, K. Principal aspects of compaction and fluid flow in mudstones — 73

Experimental studies

CLENNELL, M. B., DEWHURST, D. N., BROWN, K. M. & WESTBROOK, G. K. Permeability anisotropy of consolidated clays — 79

PETERS, M. G. & MALTMAN, A. J. Insights into the hydraulic performance of landfill-lining clays during deformation — 97

HARRINGTON, J. F. & HORSEMAN, S. T. Gas transport properties of clays and mudrocks — 107

Case Studies

INGRAM, G. M. & URAI, J. L. Top-seal leakage through faults and fractures: the role of mudrock properties — 125

SKAR, T., VAN BALEN, R. T., ARNESON. L. & CLOETINGH, S. Origin of overpressures on the Halten Terrace, offshore mid-Norway: the potential role of mechanical compaction, pressure transfer and stress — 137

DORSCH, J. & KATSUBE, T. J. Porosity characteristics of Cambrian mudrocks (Oak Ridge, East Tennessee, USA) and their implications for contaminant transport — 157

WALRAEVENS, K. & CARDENAL, J. Preferential pathways in an Eocene clay: hydrogeological and hydrogeochemical evidence — 175

Index — 187

Muds and mudstones: physical and fluid-flow properties

ANDREW C. APLIN[1], ANDREW J. FLEET[2] & JOE H. S. MACQUAKER[3]

[1] *Fossil Fuels and Environmental Geochemistry Postgraduate Institute: NRG, Drummond Building, University of Newcastle, Newcastle upon Tyne NE1 7RU, UK*
[2] *Department of Mineralogy, The Natural History Museum, Cromwell Road, London SW7 5BD, UK*
[3] *Department of Geology, University of Manchester, Oxford Road, Manchester M13 9PL, UK*

Muds and mudstones are the prime control on fluid flow in sedimentary basins and near-surface environments. As the world's commonest sediment type, they act as aquitards in sedimentary basins, restricting water flow, and they influence the development of overpressure. In petroleum systems they act as source rocks for nearly all oil and much gas, determine migration directions between source and trap in most settings, and act as seals to many reservoirs. In near surface environments, they not only control natural flow, but have also been used over the centuries to restrict leakage, most pertinently in recent times from waste disposal sites.

This book focuses on fluid flow through muds and mudstones. Such flow controls processes such as water escape from a mud during burial, upward or downward petroleum expulsion from a source-rock sequence, leakage from a petroleum reservoir, or containment of leachate in a clay-lined landfill site.

Despite the significance of muds and mudstones, their fine-grained nature means that our knowledge of their composition and properties lags behind that of other sediments. Their physical and bulk properties are poorly defined, particularly as they relate to behaviour at depth; for instance, what mudstone permeability should be applied when carrying out a particular fluid-flow modelling exercise, or under what conditions does flow through fractures dominate flow through the capillary matrix?

A search of the Science Citation Index from 1981 to 1998 revealed 13 380 articles containing the word 'mud' or 'shale' in the title, keyword or abstract; about 750 per year. 5986 articles contain 'mud' or 'shale' in the title alone. However, surprisingly few articles and books consider the key physical properties of muds, despite the volumetric importance of muds and the fact that an understanding of their properties is fundamental to a whole range of processes of importance to the petroleum, environmental and engineering industries (Table 1). In the geological literature, sedimentology, palaeontology, geochemistry, mineralogy, diagenesis and palaeogeography are dominant themes (e.g. Weaver & Beck 1971; Blatt *et al.* 1980; Potter

Table 1. *Muds and mudstones in the industrial sector*

Discipline	Process
Petroleum exploration and production	Drilling problems/performance
	Pore pressure prediction
	Vertical migration of petroleum
	Seal capacity and caprock integrity
Waste containment	Landfill liners
	Storage of nuclear and hazardous waste
	Contaminant transport
Engineering	Landslide prediction
	Foundation design
	Subsidence
	Swelling and shrinkage
Heavy clay industry	Brick and ceramic raw material

et al. 1980; Chamley 1989; Weaver 1989; O'Brien et al. 1990; Velde 1992), with distinctly less emphasis on physical and mechanical properties. Notable exceptions to this generalization are the important compilations of Rieke & Chilingarian (1974; physical properties), Bennett et al. (1991; microfabric) and Maltman (1994; deformation).

A recurring theme of the papers in this volume is thus the surprise expressed at the remarkably limited database describing the fundamental properties (e.g. thermal conductivity, permeability, strength, compressibility, mechanical behaviour, pore size distributions) of *well characterized* muds. Links between properties and sedimentary makeup remain critical because the primary sedimentological and mineralogical diversity of mudstones ensures an equally diverse range of physical properties. Without the basic data, accurate predictions and models of the processes listed in Table 1 will continue to prove elusive.

This volume seeks to take stock of our knowledge of muds and mudstones, as it relates to physical properties and fluid flow, through a series of papers which review particular topics (e.g. porosity, permeability), or discuss experimental results, or present specific case studies. Together the papers consider the physical properties of muds and mudstones from the near-surface to the deep basinal. They do not provide encyclopaedic answers but try and focus on some of the key issues which need to be resolved if fluid flow through muds and mudstones is to be understood.

Defining muds and mudstones

Both stratigraphic (Blatt 1970) and geochemical (Garrels & Mackenzie 1970) data tell us that fine-grained clastic sediments are the world's commonest sediment type, comprising more than 65% of the sediment pile. They are often described, in hand specimen, by a confusing plethora of terms (for example, clay, mud, mudstone, shale, claystone, siltstone, argillite) and qualifiers (for example, silty, clay-rich, silt-rich, shaly). Defining clay as particles finer than 1/256 mm (c. 4 µm) diameter and silt as particles with diameters between 1/256 and 1/16 mm (62.5 µm), Blatt et al. (1980) provided a useful, simplified terminology based purely on grain size. Using this classification scheme, mud is a sediment predominantly composed of clay and silt, a mudstone is a sedimentary rock composed of lithified mud and shale is a fissile mudstone.

Blatt et al. (1980) also provide a more detailed classification of fine-grained clastic sediments, which, as for most sedimentary rocks, is based on grain size and texture:

	Fissile mudstone	Non-fissile mudstone
>2/3 silt	Silt-shale	Siltstone
1/3–2/3 silt	Mud-shale	Mudstone
>2/3 clay	Clay-shale	Claystone

A classification centred on these parameters is useful as grain size is closely linked to mineralogy and the physical properties of muds (for example: Burland 1990; **Midttøme & Roaldset**; **Dewhurst et al.**, this volume). But although simple and robust, there are several difficulties with Blatt et al.'s classification. Firstly, it is relatively difficult to determine the grain size distribution of lithified muds. Chewing the sample to estimate its 'grittiness' is a common way of distinguishing silt and clay, but is scarcely quantitative! Secondly, by dividing these fine-grained sediments up into two subsets of a larger group, when the subsets themselves are difficult to distinguish, confers no obvious advantage to most geologists and thus on its own would not be adopted. Thirdly, using fissility as a descriptive parameter is fraught with uncertainty as fissility commonly develops with increased weathering. Fourthly, not taking into account the mineralogy and origin of the grains within the fine-grained sediment has the potential to confuse, particularly where fine-grained carbonate rocks are encountered.

Caught between common usage and the strict use of a term, we have chosen to use 'mud and mudstone' in their more general sense: to indicate clastic sediments and sedimentary rocks which are formed primarily from particles smaller than 1/16 mm (62.5 µm). These terms have been adopted because the name mudstone is consistent with sandstone and limestone. Furthermore, it is less ambiguous than the term 'shale', which to be effective relies partly on the degree of weathering.

Often grey and apparently homogeneous in hand specimen, the rich diversity of mudstones is only revealed by X-radiography or under the optical or electron microscope (O'Brien & Slatt 1990). These techniques reveal significant textural, grain size and compositional variability and suggest that this variability can be systematic in both temporal and spatial senses (Macquaker et al. 1998).

Most muds are deposited as an assemblage of particles ranging in diameter from less than 0.1 µm to greater than 100 µm. This three orders of magnitude range, whilst typical of muds, is unusual in other clastic sediments; for example, very few sandstones contain particles ranging in size from the very finest sand (62 μm) to cobbles (0.062 m). The mean grain size of muds is also very variable, with clay fractions ranging from 10 to 100%. For example, Picard's (1971) average composition for 751 recent muds (15% sand, 45% silt and 40% clay) derives from samples with clay fractions ranging from 0 to 95%. Deposition of distinctive muds in both space and time results in mudstone packages which are lithologically heterogeneous on scales which can be observed on photomicrographs (Potter et al. 1980; O'Brien & Slatt 1990; Bennett et al. 1991; Macquaker & Gawthorpe 1993), X-radiographs (O'Brien & Slatt 1990), wireline-log data (Bohacs & Schwalbach 1992) and even seismic data (Cartwright & Dewhurst 1998). Although generally ignored to date, heterogeneity poses important future challenges to those seeking to model the flow of fluid and heat through mudstones and the deformation of thick mudstone packages.

This volume

The papers in this volume are written by people working on both petroleum and environmental issues. Both communities take an active interest in muds and mudstones since abundant petroleum is generated from and trapped by mudstones, and because huge amounts of domestic and industrial waste are retained by muds. An example, in stark financial terms, of their significance is that mudstones cause 90% of drilling problems for the petroleum industry and cost the petroleum drilling industry around $600 million per year.

Three general themes, closely linked but studied separately within sub-disciplines of earth sciences, form the basis of the volume:

(1) transport properties, including absolute and relative permeability, diffusion rates and the role of faults, fractures and shears as fluid conduits;
(2) mechanical properties, including deformation and the development of faults and fractures;
(3) geochemical properties, in particular the colloidal properties of mudstones.

Physical properties

The first five papers of the volume consider the basic physical properties of mudstones. Porosity, probably the most fundamental property of a sediment, is reviewed by **Pearson**. Even this is subject to uncertainty! **Pearson** argues that the relevant measure of porosity depends on what process one is trying to model. He distinguishes three types of porosity.

(1) Physical porosity, which is the volume not occupied by grains.
(2) Transport porosity, which is the interconnected physical porosity and can be estimated by mercury injection porosimetry or measurements of water content. Transport porosity can be sub-divided into advective and diffusive transport porosity, which are used when describing the movement of fluids and solutes. Water diffusion porosities equal water content porosities but solute diffusion porosities are smaller because solutes cannot access the entire water content of mudstones as a result of minerals' electrical double layers and the common occurrence of nanometre-scale pores in deeply-buried mudstones.
(3) Geochemical porosity. This is considered to be the fluid volume in which geochemical reactions occur and is the value required for geochemical modelling. Based on halide data, Pearson argues that the geochemical porosity is similar to the solute diffusion porosity and is only 30–70% of other porosity values. These differences are unimportant for transport calculations but lead to uncertainties when pore water compositions are modelled.

The permeability minefield is reviewed by **Dewhurst et al.**, who consider pores, fractures and faults as possible fluid conduits. Mudstones exhibit a ten orders of magnitude permeability range with a three orders of magnitude range at a single porosity. Currently, predictions of fluid flow and pressure in sedimentary basins are severely impeded by a lack of basic porosity–permeability relationships for mudstones. Based on pore-size distributions and a sadly diminutive number of permeability data for well-characterized mudstones, there is a strong hint that much of the variation is related to grain size, with coarser-grained muds having higher permeabilities at a given porosity. Equally, the rate of loss of permeability with porosity appears to be influenced by grain size, with compaction occurring by collapse of the largest pores through which most fluid flow occurs.

As faults appear to act both as fluid conduits and barriers, under what conditions do they exhibit enhanced permeability? Flow must be episodic and is perhaps limited to periods of active fault movement, since both field and laboratory evidence shows that matrix permeabilities in muddy fault zones are very low and similar to those of the wall rocks. Experimental evidence, including that presented by **Peters & Maltman**, in this volume, suggests that faults and shear zones can be sites of preferred fluid flow, for water and perhaps petroleum, during active deformation. Dilation enhances permeability and is more likely to occur in overconsolidated (uplifted) and brittle (lithified) muds (**Ingram & Urai**).

Hydraulically formed microfractures remain a mystery. They are rarely seen in mudstones but are often inferred from the fact that observed fluid pressures do not exceed minimum stress (e.g. Düppenbecker et al. 1991). For oil petroleum companies, this is important because the probability of petroleum leakage through caprocks appears to be enhanced when fluid pressures approach the minimum stress (Gaarenstroom et al. 1993; **Ingram & Urai**). Important questions relating to issues surrounding microfractures are whether they actually occur and if so, how much fluid is lost, and over what timescale is permeability enhanced? In their paper, **Ingram & Urai** suggest that most mudstone caprocks will fail by shear, leading to enhanced permeability if the mudstones are dilated. Dilation is more likely to occur in stronger (overconsolidated and/or lithified) mudstones.

Midttøme & Roaldset's review of thermal conductivity carries similar messages to **Dewhurst et al.**'s review of permeability. The lack of basic thermal conductivity data, especially for well-characterized mudstones, means that there is no clear understanding of what controls the range of measured values. Use of inappropriate values in computer-based thermal models of sedimentary basins can lead to dramatic differences in temperature histories and thus the time at which oil and gas were generated. One example shows a 50°C variation in the modelled, present-day temperature of some upper Jurassic sediments from the northern North Sea, depending which of two different methods was used to estimate matrix conductivity. The need to standardise the technique used to measure thermal conductivity is emphasized, as is the need to understand the influence of sediment texture (mineralogy, grain size, sedimentology) on thermal conductivity.

The deformation and mechanical failure of mudstones is considered by **Petley**. Combining a new set of high mean effective stress (2–50 MPa), undrained shear experiments on London Clay with a summary of similar experiments on other mudstone datasets, he constructs a new framework for the behaviour of mudstones during undrained shear deformation. Previous work has suggested that at low mean effective stress, most undisturbed mudstones behave in a brittle manner, showing distinct peak strength before undergoing failure and strain weakening to a residual strength. In contrast, at very high mean effective stresses, mudstones tend to behave in a ductile manner, with the maintenance of peak strengths to large strains. In this paper, **Petley** suggests that a 'transitional regime' can be defined for most mudstones, in which undrained shear deformation leads to the maintenance of peak strength to a given axial strain in a manner which is similar to ductile deformation, before the initiation of strain weakening to a residual strength. Pervasive microfracturing characterizes the transition zone so that the mudstone behaves in a ductile manner on the macro-scale and in a brittle manner on the micro-scale. A critical question is then: what happens to the large-scale permeability of the mudstone along the deformation path (see **Ingram & Urai**; **Dewhurst et al.** and **Peters & Maltman**)?

The effects of chemical diagenesis on mudstone properties are highlighted by **Bjørlykke**. He reminds us that, although mudstone compaction is usually modelled as a purely mechanical effect, above 60°C time–temperature dependent *chemical* processes may also be important, as they are in sandstones. There are few quantitative data to constrain the rate of chemical compaction or its impact on permeability and overpressure development, but, as **Bjørlykke** points out, chemical compaction may proceed with a much lower dependence on effective stress. In addition, chemical processes will lithify the mud, increasing its brittleness and thus its propensity to develop dilatational fractures (see also **Ingram & Urai** and **Petley**).

Experimental studies

The next three papers in the volume are related in that they are all experimental studies of mud permeability. Although the rationale for the work was different in each case (geological, storage of radioactive waste, landfill liner), the ideas and data are readily transferable between sub-disciplines.

There are only a handful of published studies of permeability anisotropy and the way that this evolves with increasing stress. **Clennell et al.** examine the development of permeability anisotropy in well characterized kaolinite, smectite, silty clay and a remoulded, natural mud, experimentally compacted to 4 MPa. Anisotropies are less than 3 for natural clays and kaolinite, but is 8 at a porosity of 60% for pure Ca montmorillonite. These values are similar to previously published values (Tavenas et al. 1983;

Al-Tabbaa & Wood 1987; Leroueil et al. 1990), but much less than those predicted from models in which clay plates become increasingly aligned with increasing stress (up to 100; Arch & Maltman 1990). Photomicrographs suggest that compaction-driven reorientation of clay plates is not easy and that directional permeability is not simply controlled by levels of average particle orientation. A two-tier microstructural hierarchy is proposed, comprising randomly aligned particle domains and shear faces between domains. Permeability anisotropy models based on a mixing of plates, ellipses and spheres are more applicable to muds and predict anisotropy levels of 2-3. Greater anisotropies may only be possible in laminated or vertically stacked, heterogeneous mudstone sequences.

Permeability changes as a result of failure are important to all situations where fluid retention is important: oil or gas field caprocks, hazardous waste storage sites and landfill liners. **Peters & Maltman** present an experimental investigation of strain and shear-related changes in the hydraulic conductivity of natural boulder clays, which operate as landfill liners in many parts of the UK. Many of these clays are overconsolidated due to the removal of Quaternary ice. Shearing of overconsolidated clays is accommodated by dilation, with potentially increased permeability. With increasing strain, the hydraulic conductivity of an overconsolidated sample increased by less than 20% and by less than 10% at failure.

Two-phase flow in fine-grained materials is another poorly constrained process of importance to both the petroleum and radioactive waste communities. Gas generated within storage sites may increase the pressure within clay-liners and lead to its fracture, potentially compromising the long-term performance of the repository. In similar vein, migration or remigration of petroleum through thick sequences of mudstones is a common, critical, but poorly understood phenomenon, in many parts of the world (e.g. Gulf Coast, Niger Delta, North Sea Tertiary). Debate continues over the relative importance of migration through capillary networks (Leith et al. 1993), faults (Hooper 1991) and microfractures (Caillet 1993) resulting from high fluid pressures or changes in the local stress regime.

In the radioactive waste context, **Harrington & Horseman** have performed gas injection experiments on compacted bentonite (landfill liner or liner for nuclear waste repository) and a natural mudstone, the Boom Clay. Intrinsic permeabilities were around 100-500 nD for the Boom Clay and 5-10 nD for the bentonite at porosities of close to 40%. In these experiments, gas only flowed through pressure-induced cracks or microfractures, not through the intergranular pores. Gas permeabilities were similar to the intrinsic permeabilities and depended on the number of pressure-induced flow paths. From a petroleum perspective, it is intriguing to consider these results in the context of the abundant geochemical and geophysical evidence for vertical migration of gas through mudstones in sedimentary basins; debate will continue as to the relative importance of capillary and fracture flow as gas flow mechanisms in these cases.

Case studies

Caprock leakage from petroleum reservoirs is further considered by **Ingram & Urai**. Where the capillary threshold pressure of a top seal is greater than the buoyant pressure of a petroleum column in a fill-to-spill reservoir, leakage is only possible through faults or fractures. **Ingram & Urai** propose some simple but pragmatic predictive tools to risk the likelihood of top seal leakage through faults and fractures. For faults, fault density and the internal geometry/stratigraphy of the mudstone are critical because of the increased risk of leakage when relatively permeable silts/thin sands with low capillary threshold pressures and relatively high permeabilities are juxtaposed by faults.

Leakage may also occur through extensional or dilatant shear fractures. By pointing out that stronger (overconsolidated and/or cemented) mudstones are more likely to dilate than weaker mudstones, **Ingram & Urai** define strategies, based on parameters which can either be easily measured or estimated from wireline data, to evaluate the risk of dilatational fractures. One important question is exactly how muds become lithified into mudstones, since these are the processes which increase strength and the tendency to deform by brittle fracture. These processes remain poorly constrained in mudstones, in part because it is difficult to observe mineral cements except when they are obviously pore-filling.

Overpressure (that is, fluid pressures in excess of hydrostatic) develops in sedimentary basins where the rate of pressure generation by sediment loading or gas generation exceeds the rate at which it can be dissipated by fluid flow. In this context mudstones are critical in that they are the most common, low permeability unit in basins and allow the retention of high fluid pressures over geological timescales. On the other hand, the low permeability of mudstones makes it impossible to directly measure the pore pressure, which must therefore be inferred from other physical properties. In their paper, **Skar et al.** estimate pore pressures of mudstones on the Halten Terrace using soil mechanics consolidation theory to relate measured porosities to effective stress (=overburden stress − pore pressure; Hottmann & Johnson 1965). Their analysis suggests distinct variations in pore pressure which cannot be easily explained by disequilibrium compaction resulting from rapid, late Tertiary sedimentation. **Skar et al.**'s explanation is that pressure has been transferred laterally from the deep Rås Basin through the Klakk Fault Complex. At the present time, the strong pressure difference across the fault complex indicates that it is sealing. The key question, which is equally relevant to fault seal in reservoirs, is thus understanding and predicting the circumstances under which faults transmit fluid (see **Dewhurst et al.**). In this case, **Skar et al.** argue that the fault transmitted fluid and thus pressure is a result of a basinal reorientation of stress in the early Pliocene.

The last two papers concern the pore structure and flow properties of mudstones which are currently close to the earth's surface. **Dorsch & Katsube**'s interest lies in the porosity and pore-size distribution of mudstones of the Cambrian Conasuga Group, which has been investigated as a possible site for the storage of radioactive contaminants. As also shown by **Pearson**'s work, different techniques resulted in different porosity values, ranging between around 4 and 10% for mudstones that have been buried to over 4 km. Values determined by mercury injection are lowest because this technique only detects pores with diameters greater than 3 nm. Deeply buried mudstones contain significant porosity with throats smaller than 3 nm. This nanoporosity may be inaccessible to contaminants which occur as organic chelates or as colloids, which will therefore follow preferential flow pathways through larger pores.

Preferential groundwater flow pathways through the shallow buried, Tertiary Bartonian Clay in Belgium are also inferred by **Walraevens & Cardenal**, based on the inconsistency between laboratory permeability (10^{-10} m s^{-1}) tests and the distribution of hydraulic heads in the region. Hydraulic conductivities of 10^{-9} m s^{-1} gave an excellent agreement between measured and modelled hydraulic heads, thus implying the presence of preferential flowpaths. Recharge to the underlying Ledo–Paniselian aquifer occurs through the Bartonian Clay. Hydrogeochemical data are used to support the notion of localized, preferential flow pathways, which may in principle be faults, fractures or thin, interconnected silts or sands.

The relevance of these results to the petroleum sector is striking since, as we have already pointed out, that industry is still struggling to define the pathways by which petroleum migrates through kilometre-thick mudstone sequences (e.g. Goff 1984). Preferential pathways are often inferred (e.g. Hooper 1991; Roberts & Nunn 1995; Larter et al. 1996) but there is little consensus as to the nature of the pathways (faults, fractures, silts) and the circumstances under which they are able to conduct fluid.

Future directions

The papers in this volume provide a brief snapshot of some of the current research on fine-grained sediments. By highlighting defects in our knowledge, it provides pointers to the future. Even if the coming millennium is not ultimately remembered as the millennium of mud, in its first decade the following areas seem likely to be amongst those in which progress will be made.

(1) *Gathering of basic data for the physical properties of well-characterized mudstones, including compressibility, strength, absolute and relative permeability, thermal conductivity, mechanical behaviour and petrophysical properties.* If apparently mundane, this task is critical as it will underpin all efforts to provide a predictive understanding of the environmental behaviour of mudstones in a range of settings and on both human (e.g. drilling) and geological timescales. Proper account should of course be taken of the sedimentological diversity exhibited by mudstones.

(2) *Mechanisms of single and multiphase fluid flow*. In addition to the continuing need for single phase permeability data, it will also be important to constrain the relative importance of capillaries, faults and fractures on flow both through and across mudstone sequences. Experimental studies of the hydromechanical behaviour of mudstones, especially at high fluid pressures where hydrofracturing may occur and at which permeability may be strongly related to stress will be important. Establishing ways in which data gathered from short-term laboratory experiments can be confidently extrapolated to explain processes on geological timescales will, as always, be a point of discussion. All these ideas are also relevant to the placing of constraints on the migration mechanisms of gaseous and liquid petroleum, a problem of real importance for both petroleum and environmental scientists. Again, the relative importance of fracture v. capillary flow is critical, as is the definition of wettability. Although some ideas can be imported from the extensive studies of

multiphase flow in petroleum reservoirs, the low permeability and sub micron pore size of mudstones ensure that this is an area where progress will require the collaboration of geoscientists with physicists and chemists.

(3) *Definition and visualization of three dimensional mudstone sequences.* Into the future, the fundamental aim must be to advance the study of mudstones to a similar point to that enjoyed by sandstones and carbonates today. Techniques initially developed for the exploration and production of petroleum can now be exploited as a tool for defining and interpreting mudstone sequences. Wireline well log and both 2D and 3D seismic data can be used to help define the anatomy, petrophysical and geochemical properties of mudstone sequences. In turn, these data form the basis for placing the mudstones, and their specific characteristics, into a chronostratigraphic, sequence stratigraphic and depositional framework. Ultimately, stratigraphic data may be tied to the tectonic, weathering, transport and depositional processes which impose the grain size and mineralogical properties which exert a fundamental control on the muds' physical behaviour upon burial.

(4) *Computer-based modelling.* Increased computing power means that three dimensional geological data is becoming easier and easier to assimilate and display. For mudstones, this implies that compaction will be treated in terms of a tensorial rheology; models already exist in civil engineering but have yet to be used successfully to explain geological compaction in sedimentary basins (Pouya *et al.* 1998). With appropriate calibration to wireline log and physical parameters, 3D seismic data can further define the internal anatomy of mudstone sequences and the occurrence of internal faults and fractures (Cartwright & Dewhurst 1998; **Ingram & Urai**). Flow simulators will be increasingly combined with models describing the chemical interactions between water, organic fluids and rock minerals. The high specific surface area of mudstones implies that particular attention should be given to adsorption and ion exchange reactions.

(5) *Mechanical properties.* A more unified understanding of the mechanical (compaction, fracture) properties of mudstones is likely (see Maltman 1994 and **Petley**). Historically, compaction has been treated primarily as an elastoplastic, mechanical process. However, it is likely that irreversible, temperature-dependent chemical processes are also operating at depths greater than approximately 2 km (**Bjørlykke**). The manner in which these chemical changes lithify, and thus fundamentally alter the strengthen the mechanical properties of mudstones is poorly understood but their importance lie in the fact that they will strongly influence the fracture, fluid flow and sealing characteristics of mudstones. Equally important will be to define the links between failure mechanisms and the basic mechanical properties of the full range of mudstone types.

References

AL-TABBAA, A. & WOOD, D. M. 1987. Some measurements of the permeability of kaolin. *Géotechnique*, **37**, 499–503.

ARCH, J. & MALTMAN, A. J. (1990). Anisotropic permeability and tortuosity in deformed wet sediments. *Journal of Geophysical Research*, **95**, 9035–9047.

BENNETT, R. H., BRYANT, W. R. & HULBERT, M.H. (eds) 1991. *The Microstructure of Fine-grained Sediments, from Mud to Shale.* Springer Verlag, New York.

BLATT, H. 1970. Determination of mean sediment thickness in the crust: a sedimentological method. *Geological Society of America Bulletin*, **81**, 255–262.

——, MIDDLETON, G. V. & MURRAY, R. C. 1980. *Origin of Sedimentary Rocks,* 2nd Edition, Prentice-Hall, Englewood Cliffs, NJ.

BOHACS, K. M. & SCHWALBACH, J. R. 1992. Sequence Stratigraphy in Fine-Grained Rocks: With Special Reference to the Monterey Formation. *In*: SCHWALBACH, J. R. & BOHACS, K. M. (eds) *Sequence Stratigraphy in Fine-Grained Rocks: Examples from the Monterey Formation.* SEPM Field Guide, 7–20.

BURLAND, J. B. 1990. On the compressibility and shear strength of natural clays. *Géotechnique*, **40**, 329–378.

CAILLET, G. 1993. The caprock of the Snorre Field, Norway: a possible leakage by hydraulic fracturing. *Marine and Petroleum Geology*, **10**, 42–50.

CARTWRIGHT, J. A. & DEWHURST, D. N. 1998. Layerbound compaction faults in fine-grained sediments. *Geological Society of America Bulletin*, **110**, 1242–1257.

CHAMLEY, H. 1989. *Clay Sedimentology.* Springer-Verlag, Berlin.

DÜPPENBECKER, S. J., DOHMEN, L. & WELTE, D. H. 1991. Numerical modelling of petroleum expulsion in two areas of the Lower Saxony Basin, northern Germany. *In*: ENGLAND, W. A. & FLEET, A. J. (eds) *Petroleum Migration.* Geological Society, London, Special Publications, **59**, 47–64.

GAARENSTROOM, L., TROMP, R. A. J., DE JONG, M. C. & BRANDENBURG, A. M. 1993. Overpressures in the Central North Sea: implications for trap integrity and drilling safety. *In*: PARKER, J. R. (ed.) *Petroleum Geology of Northwest Europe: Proceedings of the 4th Conference.* Geological Society of London, 1305–1313.

GARRELS, R. M. & MACKENZIE, F. T. 1970. *Evolution of Sedimentary Rocks.* W.W. Norton and Co.

GOFF, J. C. 1984. Hydocarbon generation and migration from Jurassic source rocks in the East Shetland Basin and Viking Graben of the northern North Sea. *Journal of the Geological Society, London*, **140**, 445–474.

HOOPER, E. C. D. 1991. Fluid migration along growth faults in compacting sediments. *Journal of Petroleum Geology*, **14**, 161–180.

HOTTMANN, C. E. & JOHNSON, R. K. 1965. Estimation of formation pressures from log-derived shale properties. *Journal of Petroleum Technology*, June, 717–722.

LARTER, S. R., TAYLOR, P., CHEN, M., BOWLER, B., RINGROSE, P. & HORSTAD, I. 1996. Secondary migration — visualising the invisible — what can geochemistry potentially do? *In*: GLENNIE, K. & HURST, A. (eds) *AD1995: NW Europe's Hydrocarbon Industry*. Geological Society, London, 137–143.

LEITH, T. L., KAARSTAD, I., CONNAN, J., PIERRON, J. & CAILLET, G. 1993. Recognition of caprock leakage in the Snorre Field, Norwegian North Sea. *Marine and Petroleum Geology*, **10**, 29–41.

LEROUEIL, S., BOUCLIN, G., TAVENAS, F., BERGERON, I. & LA ROCHELLE, P. 1990. Permeability anisotropy of natural clays as a function of strain. *Canadian Geotechnical Journal*, **27**, 568–579.

MACQUAKER, J. H. S. & GAWTHORPE R. L. 1993. Mudstone lithofacies in the Kimmeridge Clay Formation, Wessex Basin: Implications for the origin and controls on the distribution of mudstones. *Journal of Sedimentary Petrology*, **63**, 1129–1143.

——, ——, TAYLOR, K. G. & OATES, M. J. 1998. Heterogeneity, stacking patterns and sequence stratigraphic interpretation in distal mudstone successions: examples from the Kimmeridge Clay Formation, U.K. *In*: SCHIEBER, J., ZIMMERLE, W. & SETHI, P. (eds) *Recent Progress in Shale Research*. Schweizerbartísche Verlagsbuchhandlung, Stuttgart, 163–186.

MALTMAN, A. (ed.) 1994. *The Geological Deformation of Sediments*. Chapman and Hall, London.

O'BRIEN, N. R. & SLATT, R. M. 1990. *Argillaceous Rock Atlas*. Springer Verlag, New York.

PICARD, M. D. 1971. Classification of fine-grained sedimentary rocks. *Journal of Sedimentary Petrology*, **41**, 179–195.

POTTER, P. E., MAYNARD, J. B. & PRYOR, W. A. 1980. *Sedimentology of Shale*. Springer-Verlag, New York.

POUYA, A., DJERAN-MAIGRE, I., LAMOUREUX-VAR, V. & GRUNBERGER, D. 1998. Mechanical behaviour of fine grained sediments: experimental compaction and three-dimensional constitutive model. *Marine and Petroleum Geology*, **15**, 129–143.

RIEKE, H. H. & CHILINGARIAN, G. V. 1974. *Compaction of Argillaceous Sediments*. Developments in Sedimentology **16**.

ROBERTS, S. J. & NUNN, J. A. 1995. Episodic fluid expulsion from geopressured sediments. *Marine and Petroleum Geology*, **12**, 195–204.

TAVENAS, F., JEAN, P., LEBLOND, P. & LEROUEIL, S. 1983. The permeability of natural soft clays. Part 2. Permeability characteristics. *Canadian Geotechnical Journal*, **20**, 645–660.

VELDE, B. 1992. *Introduction to Clay Minerals: Chemistry, Origins, Uses, and Environmental Significance*. Chapman & Hall, London.

WEAVER, C. E. 1989. *Clays, Muds, and Shales*. Developments in Sedimentology **44**.

WEAVER, C. E. & BECK, K. C. 1971. Clay water diagenesis during burial: how mud becomes gneiss. Geological Society of America Special Paper **134**.

What is the porosity of a mudrock?

F. J. PEARSON

Waste Management Laboratory, Paul Scherrer Institute, CH-5232 Villigen PSI, Switzerland
Present address: Ground-Water Geochemistry, 411 East Front Street, New Bern,
NC 28560, USA

Abstract: Several types of porosity (total or physical, advective flow, advective and diffusive transport, and geochemical) can be distinguished in a mudrock. Each may have a different value. Physical porosity is the volume not occupied by mineral grains. Interconnected physical porosity can be found using injection techniques and water-content measurements. Advective and diffusive flow and transport porosities are used when describing the movement of fluids and solutes. Advective porosity relates the average linear velocity of a fluid to its Darcy flux (velocity), and in mudrock is relevant principally to fractures. Diffusive porosity relates diffusion in rock to diffusion in pure water and depends on the diffusing substance and the medium. In mudrock, water diffusion porosities equal water content porosities but solute diffusion porosities are smaller because solutes do not have access to the entire water content of a rock. Inaccessible water may be in pores too small to admit solutes, or present as interlayer or surface-layer water. Geochemical porosity is required to model pore-water composition and reactive transport. It represents the fluid volume in which reactions occur, and for mudrocks, seems equivalent to solute-diffusion porosity. Review of studies of several mudrocks shows halide geochemical porosities from 0.3 to 0.7 of other porosity values. Uncertainties of this size are unimportant for transport calculations, but lead to significant uncertainties in modelled pore-water compositions. To make acceptably accurate calculations of water–rock reactions and transport in mudrock, it is important to define the geochemical porosity.

The term 'porosity' has different meanings in various disciplines, and numerical values of the property it refers to may also differ depending on the technique used to measure it. To petrographers and others interested in the structure of rock and soil, porosity describes that part of a material that is not occupied by mineral grains, and it is commonly measured by density difference, injection, or fluid loss methods. Hydrologists and petroleum scientists interested in the movement of fluids and solutes through rock are concerned with porosity values as parameters in their mathematical descriptions of transport velocities and fluxes. Transport porosities are usually measured in transport experiments. It is widely recognized that the transport porosity of hydrologists is commonly less than the void volume, or physical porosity, and the term *effective porosity* is often used to distinguish the porosity used in transport modelling from the *physical* or *total porosity* (Freeze & Cherry 1979; Domenico & Schwartz 1990).

Geochemical and reactive transport models are commonly written using solution concentrations expressed as molality (mol kg^{-1} H$_2$O) or molarity (mol l^{-1} solution). It is convenient also to express concentrations of solid-phase reactants and products per mass of H$_2$O or volume of solution. To do this requires knowledge of a third type of porosity, which can be called *geochemical porosity*. Geochemical porosity is the proportion of the total volume of a soil or rock occupied by the pore fluid in which occur the reactions of interest.

Differences among the values of physical, transport, and geochemical porosities are relatively small in coarse-grained rock. In such rock, with void space well interconnected by pores that are large relative to the size of fluid molecules and dissolved substances, and composed of minerals with relatively non-reactive surfaces, the observable, physically measurable porosity will have values similar to porosities determined from solute transport experiments. Both will be the same as the geochemical porosity because virtually all the pore water is available as a medium for chemical reactions. For many rocks in which solute transport and reactive transport studies are carried out, differences among the values of the three types of porosity are too small to be of much concern.

From: APLIN, A. C., FLEET, A. J. & MACQUAKER, J. H. S. (eds) *Muds and Mudstones:*
Physical and Fluid Flow Properties. Geological Society, London, Special Publications, **158**,
9–21. 1-86239-030-4/99/$15.00 © The Geological Society of London 1999.

Some rocks include porosity that is not connected with the primary pore structure. The presence of such isolated pores, as well as of fluid inclusions in minerals, makes it necessary to consider total and connected physical porosity separately. Furthermore, all connected porosity is not identical, for whereas some pores are connected so that flow carrying solutes can take place through them, other pores may be connected only at one end. No fluid flux or advective solute transport occurs into these dead-end pores, but solutes can enter or leave them by diffusion. Therefore, it may be necessary to distinguish between types of connected porosity that permit advective flow and transport and types that permit only diffusive transport (Norton & Knapp 1977).

In fine-grained rocks with small pores such as mudrock, a certain amount of the water or other pore fluid will be ordered or structured by its association with mineral surfaces and clay interlayer cations. In addition, the sizes of some pore throats may become so small that solutes cannot enter them. This gives rise to the osmotic membrane properties of some mudrocks. Pore water that is structured or in small pores will not be accessible to solutes for either transport or chemical reactions. Thus both the geochemical and transport porosities of solutes will be lower than the physical porosity.

The permeability of mudrocks tends to be low and to decrease with decreasing porosity (Neuzil 1994). Because of this low permeability, transport in mudrocks is principally by diffusion. Advective transport, if it occurs at all, will generally be through localized regions of higher permeability such as fractures, fracture zones, or coarse-grained horizons within the body of the mudrock. The terms fracture and matrix porosity may be used in connection with discontinuous or dual porosity mudrocks of this type. Descriptions of types of fracture porosities (distinctions between advective and transport fracture porosities, for example) are based on idealized, conceptual models of fracture geometry (e.g. Gelhar 1987; Ando & Vomvoris, pers. comm.). The distinctions among porosity types discussed in this paper refer primarily to matrix porosity.

The principal topic of this paper will be geochemical porosity, because it is not a widely applied concept. The relationship of geochemical porosity to other types of porosity will be discussed, on the basis of a review of selected, published sets of measurements of mudrock porosities. This discussion is preceded by a review of definitions and measures of all types of porosity.

Physical porosity

The most straightforward of the three porosity types is physical porosity n_{phys}, which is simply the ratio of void volume to total volume. Total physical porosity can be calculated from the bulk density of a dry sample, $\rho_{bulk,dry}$, and its grain density, ρ_{grain}:

$$n_{phys} = \frac{V_{void}}{V_{total}} = 1 - \frac{\rho_{bulk,dry}}{\rho_{grain}}. \qquad (1)$$

To ensure that isolated pores and fluid inclusions are included in the total physical porosity value, the grain size on which grain density is measured must be smaller than the smallest pore size (Norton & Knapp 1977).

Sample impregnation is one technique used to measure the volume of connected pores. The distribution of dye-containing material infused into a sample can be examined optically, for example, to discern the pattern and quantity of pores with sizes down to the visible limit ($c.\,0.5\,\mu m$). Connected porosity can also be measured using the change in weight of a sample as mercury is injected into it. Mercury injection (mercury porosimetry) results also reflect the distribution of pores of various sizes, on the principle that the sizes of the pore throats through which mercury will pass can be related to the pressure with which the mercury is forced into the sample (see Dorsch & Katsube, this volume).

Values for connected porosity are often based on measurements of the water contents of saturated rock. Water contents, WC, are found from the relative masses of a sample when water saturated, m_{satd}, and when dry, m_{dry}:

$$WC = \frac{m_{satd} - m_{dry}}{m_{satd}} = 1 - \frac{m_{dry}}{m_{satd}}. \qquad (2)$$

Water-content porosities, n_{WC}, are calculated from measured water contents and sample densities:

$$n_{WC} = WC \frac{\rho_{bulk,satd}}{\rho_{water}}$$
$$= \frac{WC\rho_{grain}}{WC\rho_{grain} + (1 - WC)\rho_{fluid}}. \qquad (3)$$

Hydrologists use the term volumetric moisture content, or simply moisture content, θ, to refer to the ratio of water volume to total sample volume in both saturated and unsaturated rock and soils (Freeze & Cherry 1979; Domenico & Schwartz 1990). In the notation of this paper, the water-content porosity, n_{WC}, equals the moisture content of a sample at saturation, θ_{satd}.

For a mudrock, the amount of water lost depends on the drying conditions, because these conditions will determine the amount of structured water, that is, surface sorbed or interlayer water, that is removed from the sample. Because impregnation and water-content techniques measure connected rather than total porosity,

$$n_{WC} \leqslant n_{phys}.$$

As the samples of interest become more fine grained and clay-mineral rich, it is to be expected that porosity values from mercury-porosimetry and water-content measurements may diverge. For example, the pressures required to inject mercury into small pores may damage the fabric of mudrocks (Horseman *et al.* 1996). Also, as a matter of definition, it is not clear whether water removed from clay interlayer positions, and commonly part of water-loss measurements, should be included in the definition of physical porosity as a measure of the volume not occupied by mineral grains.

Transport porosity

Transport porosity values are used to relate velocities of non-reactive tracers in a fluid to the flux of that fluid, and, together with other variables, to relate the rates of diffusion of substances in a porous medium to their rates of diffusion in free water.

Transport occurs through pores, so that the maximum size of a substance being transported through a pore must be less than the minimum throat size of that pore. To reflect this fact explicitly, transport porosities in this paper are written with the superscript i, referring to the substance being transported.

Flow and advection porosity

The flux of a fluid moving in a rock is specified as volume per time and cross-sectional area. This flux has units of velocity (length/time) and in groundwater hydrology is referred to as the Darcy velocity. The average linear velocity of fluid moving in a porous medium equals the velocity of a perfect tracer in the system. The Darcy flux, V_{Darcy}, and average linear velocity, V^i, of a tracer, i, are related by the advective transport porosity, n_{adv}^i:

$$n_{adv}^i = \frac{V_{Darcy}}{V^i}. \tag{4}$$

Fluid flux (Darcy velocity) is defined with respect to the total (grain plus void space) cross-sectional area of the medium, although fluid moves only through the void space. Thus, the ratio of Darcy to linear flow velocity is the ratio of the area of flow to the total area. In a microscopic sense, the ratio of void to total area can vary across the thickness of a solid, but in the macroscopic sense, in which the equations of flow and transport are applied to porous media, the ratio of the cross-sectional area of flow to total area will be the same as the ratio of the volume through which flow occurs to the total volume.

Advective flow and transport take place only through connected pores. Thus, advective transport porosity does not include isolated pores, or pores open only at one end to the through-flowing, connected pore system (so-called dead-end pores). In general,

$$n_{adv}^i \leqslant n_{WC} \leqslant n_{phys}.$$

The term *effective porosity* occurs frequently in the hydrological literature. Most often it refers to what is here called *advection* or *flow porosity*, which is the ratio of Darcy velocity to average linear velocity in a porous medium. However, it sometimes also includes the diffusion porosity.

The advective porosity used in modelling flow in regional mudrock systems can be orders of magnitude smaller than the physical or water content porosity. Because of the generally very low permeability of mudrock (e.g. Neuzil 1994), advective flow through such units on a regional scale takes place only in fractures or other zones of higher permeability, if it occurs at all. If the regional geometry of such zones is not known, the use of discrete-fracture models is precluded. Thus, equivalent porous medium models are commonly used to describe regional flow. To be correct, average linear flow velocities determined from equivalent porous medium models must equal the velocities in the fractures or higher-permeability zones in which the flow is actually concentrated. For this to be the case, the porosity of the fracture zones must be distributed over the entire equivalent porous medium volume of the model.

For example, let us consider a regional mudrock system in which advective flow is concentrated in fracture zones 3 cm wide that occur on average every 10 m. The flow porosity of these zones is 0.02, on the basis of petrographic observations and injection porosimetry. If the flow through these fracture zones is modelled as if it were distributed over the entire rock volume, the flow porosity of 0.02 in these zones must also be distributed over the entire volume, that is,

$$0.02 \frac{3 \times 10^{-2} \text{ m}}{10 \text{ m}} = 6 \times 10^{-5}.$$

This is in sharp contrast to the porosity corresponding to the water content of a mudrock, which is typically in the range of 0.05–0.5.

Diffusion porosity

Diffusion of a substance through a porous medium can be described by equations analogous to Fick's laws of diffusion through free water. One-dimensional diffusive flux F^i through a porous medium is described by Fick's first law:

$$F^i = -D^i_{\text{eff}} \frac{\partial C^i}{\partial x} \quad (5)$$

where D^i_{eff} is the effective diffusion coefficient for substance i in the medium, and $\partial C^i/\partial x$ is its concentration gradient. In general, the diffusion-controlling properties of a porous medium will not be isotropic, so in experimental studies of mudrock diffusion, properties parallel and perpendicular to the bedding are often measured separately to account for the directional nature of D^i_{eff}. Here, diffusion will be written in one dimension for simplicity, as if the medium were isotropic.

Free water diffusion coefficients, D^i_0, for many substances are well known (e.g. Li & Gregory 1974), so it is convenient to express effective diffusion coefficients in terms of the free water diffusion coefficients and properties of the medium itself. Conventionally, this is done through an expression such as

$$D^i_{\text{eff}} = n^i_{\text{diff-area}} \left(\frac{\chi}{\tau^2}\right) D^i_0 \quad (6)$$

in which χ/τ^2 and $n^i_{\text{diff-area}}$ are determined by the properties of the medium and the diffusing substance.

The variables χ and τ^2 are called the constrictivity and tortuosity. They represent concepts that the narrowest part of any pore will control the flux through it, and that the tortuous flow path through a pore will be longer than the diffusion distance in free water. Strictly speaking, χ and τ^2 are defined in terms of idealizations of the microscopic geometry of the medium. Macroscopically, the variable accessible to experimental measurement is the quotient χ/τ^2, as described below.

$n^i_{\text{diff-area}}$ in equation (6) reflects the fact that the area across which diffusion occurs in a porous medium is smaller than the area across which it occurs in free water. Thus, it is analogous to advective transport porosity, n^i_{adv}.

Combining equation (6) with a statement of mass conservation yields an equation analogous to Fick's second law:

$$D^i_{\text{app}} \frac{\partial^2 C}{\partial x^2} = \frac{\partial C}{\partial t}. \quad (7)$$

D^i_{app} is often called the apparent diffusion coefficient. It is related to the effective diffusion coefficient D^i_{eff} by

$$D^i_{\text{app}} = \frac{D^i_{\text{eff}}}{\alpha^i} \quad (8)$$

so that equation (7) can also be written

$$D^i_{\text{eff}} \frac{\partial^2 C}{\partial x^2} = \alpha^i \frac{\partial C}{\partial t}. \quad (9)$$

α^i represents the volumetric capacity of the medium for substance i and is written

$$\alpha^i = n^i_{\text{diff-vol}} R^i. \quad (10)$$

$n^i_{\text{diff-vol}}$ is the diffusion-accessible porosity, which is the volume of fluid into which diffusion occurs, relative to the total volume of the medium. R^i is the retardation factor, which represents solute–medium interactions and is written

$$R^i = 1 + \frac{\rho_{\text{bulk}}}{n^i_{\text{diff-vol}}} K^i. \quad (11)$$

K^i is the solid–solution distribution ratio, with units of $(\text{mass}^i/\text{mass}_{\text{bulk}}) \times (\text{volume}_{\text{solution}}/\text{mass}^i)$, and the other terms are as before. K^i represents all reactions between substance i in solution and in the matrix solid, so it may well be dependent on C^i.

The distinction between $n^i_{\text{diff-area}}$ and $n^i_{\text{diff-vol}}$ has been shown to be important in a medium such as a fractured rock in which the porosity comprises both a through-transport porosity and a significant volume of dead-end pores. From laboratory diffusion studies using non-retarded tracers, for example, Bradbury & Green (1985) showed ratios of $n^i_{\text{diff-area}}$ to $n^i_{\text{diff-vol}}$ from 0.05 to 0.3 and from 0.1 to 0.4 in two samples of granite.

In diffusion studies on material such as mudrock, in which the quantity of dead-end pores is likely to be minimal, it is customary not to distinguish between diffusion-area and diffusion-accessible porosity, but to refer only to a single diffusion porosity, n^i_{diff} (e.g. Bourke et al. 1993; De Cannière et al. 1996).

If $n^i_{\text{diff-area}}$ equals $n^i_{\text{diff-vol}}$, equations (6), (8), and (10) can be combined to give the following expression for the apparent diffusion coefficient:

$$D^i_{\text{app}} = \frac{\left(\frac{\chi}{\tau^2}\right) D^i_0}{R^i}. \quad (12)$$

This shows that the apparent diffusion coefficient of a substance that is not retarded ($K^i = 0$; $R^i = 1$) is independent of porosity. Thus, from measured apparent diffusion coefficients and known values of free water diffusion coefficients, values of χ/τ^2, the factor dependent on sample pore geometry, can be found.

Several techniques can be used to measure diffusion transport factors. The *through-diffusion* method allows determination of both D^i_{eff} and α^i in equation (9). A number of configurations of samples and diffusing solutions are possible. In one of them, described and used by Bradbury & Green (1985) and Bourke et al. (1993), for example, a wafer of the sample is sandwiched between two reservoirs of fluid. Both reservoirs are maintained at the same pressure to eliminate advective transport, and the solutions in each are isotonic so that no osmotic fluxes occur. The substance being studied is present in one reservoir at a virtually constant concentration throughout the experiment, but is absent from the second reservoir and from the pore fluid of the sample at the start of the experiment. With the following experimental boundary conditions:

at $x = 0$, $\quad C^i_{\text{in}}(t) = $ constant
at $x = L$, $\quad C^i_{\text{out}}(t) = 0 \quad$ at $t = 0$
and $\quad C^i_{\text{out}}(t) \ll C^i_{\text{in}} \quad$ at $t = t$

and the condition that α^i is independent of C^i, a solution to equation (9) can be written (Lever 1986):

$$\frac{Q^i}{ALC_{\text{in}}} = \frac{D^i_{\text{eff}} t}{L^2} - \frac{\alpha^i}{6}$$
$$- \frac{2\alpha^i}{\pi^2} \sum_{j=1}^{\infty} \frac{(-1)^j}{j^2} \exp\left(-\frac{D^i_{\text{eff}} j^2 \pi^2 t}{L^2 \alpha^i}\right) \quad (13)$$

where A and L are the cross-sectional area and thickness of the sample, Q^i is the total amount of i diffused through the sample at time t and equals $C^i_{\text{out}} V_{\text{out}}$, where V_{out} is the fluid volume of the second reservoir.

When t is large, equation (13) becomes asymptotic to

$$\frac{C_{\text{out}}(t)}{C_{\text{in}}} = \frac{A}{V_{\text{out}} L} D^i_{\text{eff}} t - \frac{AL}{6 V_{\text{out}}} \alpha^i. \quad (14)$$

Thus, if experimental data are plotted as $C_{\text{out}}(t)/C_{\text{in}}$ v. t, the slope of the line asymptotic to the late-time data will lead to a value of D^i_{eff}, and the intercept of the line will lead to a value for α^i. As equation (10) shows, if non-retarded tracers are used, α^i equals $n^i_{\text{diff-vol}}$.

Bourke et al. (1993) described a series of experiments on the London Clay, an Eocene marine deposit with less than 30–50% of non-clay minerals, that illustrate aspects of diffusion porosity common to mudrocks. Diffusion experiments were carried out using three tracers, deuterium-enriched water, (HDO), tritium-enriched water, (HTO), and iodide, (I$^-$), in a diffusion cell designed to meet the conditions required to make equations (13) and (14) applicable (Bourke et al. 1993, fig. 13.1). Figure 1 is an example of through-diffusion data from this study. The lines represent fits of equation (13) to the data with the parameters shown in the figure.

Table 1 is a summary of the results of these experiments. The effective diffusion coefficients (called intrinsic diffusion coefficients D_i by Bourke et al.) are based on fits of equations (13) and (14) to the experimental data, as illustrated in Fig. 1. The values for both HDO and HTO are about the same and are about twice that of I$^-$.

The through-diffusion porosities are about 0.6 for both HDO and HTO. This is the same as the water-content porosity of the samples, and indicates that diffusing HDO and HTO have

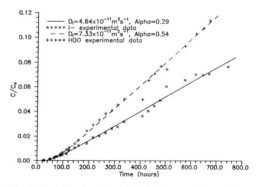

Fig. 1. Results of a through-diffusion experiment. In this figure, D_i is equivalent to the effective diffusion coefficient of this paper (Bourke et al. 1993, fig. 13.6; reprinted with permission).

Table 1. *Summary of effective diffusion coefficient values and porosities measured by various techniques (Bourke et al. 1993, table 13.3 and fig. 13.7)*

Tracer	Number of samples	D(eff) (m² s⁻¹)	Porosity			Number of in-diffusion samples
			Through-diffusion	Out-diffusion	In-diffusion	
HDO	8	7.6 ± 0.9E-11	0.61 ± 0.05	0.59 ± 0.05	0.48	3
HTO	4	7.6 ± 0.5E-11	0.56 ± 0.03	0.62 ± 0.02		
I	3	3.9 ± 0.6E-11	0.21 ± 0.06	0.32 ± 0.06	0.24–0.14	2

access to all the water in the sample that can be extracted by drying. This includes not only water in open pores, but structured water in sorbed surface layers and interlayer water, to the extent that it is removed by drying.

The through-diffusion porosity for I⁻ is 0.2, about one-third of the HDO, HTO and water content porosity. This indicates that I⁻ has access to only one-third of the sample water, presumably because its size and negative charge exclude it from very small pores and from the vicinity of negatively charged mineral surfaces.

With measured effective-diffusion coefficients and porosity values, and literature values for the free-water diffusion coefficients for H₂O and I⁻, values for χ/τ^2 can be found. The self-diffusion (tracer-diffusion) coefficient for H₂O is 2.5×10^{-9} m² s⁻¹ (Trappeniers et al. 1965) and the free-water diffusion coefficient for I⁻ is 2.0×10^{-9} m² s⁻¹ (Li & Gregory 1974). These values and the results from Table 1 give χ/τ^2 values of 0.05 for H₂O and 0.10 for I⁻ in the London Clay.

The conclusion that all of the water contained in a mudrock may not be accessible to diffusing solutes, i.e. that $n_{\text{diff}}^{\text{I}^-} < n_{\text{diff}}^{\text{H}_2\text{O}}$, can also be reached from geometric arguments. For example, Barone et al. (1990) studied chloride diffusion in a shale with a relatively low water-content porosity (0.11). On the basis of pore-size measurements and consideration of the size of the chloride ion, they concluded that at least 75%, but not all of the pore water, would be accessible to chloride diffusion.

Another method used to measure the diffusion porosity is the *out-diffusion* technique. In this technique, a mass of sample is disaggregated and leached with a volume of pure water, and the concentrations of solutes of interest in the leach solutions are measured. These concentrations are used to find the concentrations of solutes per mass sample:

$$C_r^i = C_l^i \frac{V_l}{m_r} \quad (15)$$

where C_r^i is the mass of solute i per mass of bulk sample, C_l^i is the mass of solute i per volume of leach solution and V_l/m_r is the volume of leach solution per mass of bulk sample leached.

If the concentration of the solute in the pore fluid is known, along with the sample density, ρ, the volume of solute-containing fluid per volume of rock can be found:

$$n_{\text{out-diff}}^i = C_r^i \frac{1}{C_{\text{pf}}^i} \rho_{\text{bulk}} \quad (16)$$

where C_{pf}^i is the solute concentration per volume pore fluid in the sample, and $n_{\text{out-diff}}^i$ is the out-diffusion porosity. When a through-diffusion experiment is at steady state (e.g. along the linear part of the graph of Fig. 1), the concentration of the solute in the sample pore fluid, C_{pf}^i, is the mean of C_{in}^i and C_{out}^i.

Table 1 includes porosities measured by Bourke et al. (1993) using the out-diffusion method on the same samples of London Clay as used for their through-diffusion measurements. For HDO and HTO, the out-diffusion porosities are about 0.6, the same as the through-diffusion and water-content porosities. The mean value of the I⁻ porosity determined by out-diffusion is 0.3. This is higher than the mean value from the through-diffusion measurements, although the two data sets overlap.

Bourke et al. (1993) also used an *in-diffusion* method to establish the pore fluid concentrations, C_{pf}^i, required in equation (16) to measure diffusion porosity by the out-diffusion method. As shown in Table 1, these experiments led to porosities of about 0.5 for HDO, and 0.14 and 0.24 for I⁻. Because of uncertainty in the in-diffusion results (Bourke et al. 1993, fig. 13.7), these results are probably not significantly different from the through-diffusion and out-diffusion results.

The results of the study by Bourke et al. (1993) support certain conclusions about the relative values of various types of porosity in mudrocks that can be drawn from general considerations and many other studies. For the diffusion of water molecules themselves

$$n_{adv}^{H_2O} < n_{diff}^{H_2O} \approx n_{WC} \leqslant n_{phys}$$

whereas for the diffusion of solutes

$$n_{adv}^i < n_{diff}^i < n_{WC} \leqslant n_{phys}.$$

Geochemical porosity

The need to consider a third type of porosity, geochemical porosity, arises from the requirements of geochemical and solute transport modelling. One example of this requirement can be found in the equation for the conservation of mass on which models of reactive transport are based (Domenico & Schwartz 1990; Steefel & MacQuarrie 1996):

mass inflow rate − mass outflow rate
± mass production rate
= change in mass storage with time.

This equation describes the change in mass within a unit volume of a medium. The actual transport of mass into and out of the volume occurs only in the fluid phase, and is expressed as fluid concentrations. The change in mass within the volume per unit time is related to changing fluid concentrations through

$$\Delta C_{medium} = \Delta C_{fluid} n_{geochem} \qquad (17)$$

where ΔC_{medium} is the change of concentration per volume of the medium, ΔC_{fluid} is the change of concentration per volume of solution, and $n_{geochem}$, the geochemical porosity, is the ratio of the volume of fluid in which transport and water–rock reactions occur to the total volume.

The geochemical porosity of a mudrock will certainly be less than the physical porosity because the latter may include pores and fluid inclusions that are completely isolated from the system of connected pores through which transport occurs. Diffusive and, possibly, advective transport move mass, so it seems reasonable that the geochemical porosity will be similar to the advection porosity in systems in which transport is principally by advection, and similar to the diffusion porosity in systems in which transport is principally by diffusion. Mudrocks are more likely to be of the latter type than of the former,

so for mudrocks the relative values of the various types of porosity will be

$$n_{geochem} \approx n_{diff} \leqslant n_{WC} < n_{phys}.$$

Geochemical porosity is also needed to characterize the chemistry of pore water in poorly permeable material such as mudrock. Such material generally yields little or no water to boreholes or piezometers, and indurated mudrocks (argillites) also may not yield water to high-pressure squeezing. It is possible to determine the chemistry of pore water in such rocks by modelling equilibrium reactions between the water and reactive formation minerals. Such calculations require not only the identities of reactive minerals that are present, such as carbonates and clays, but also the specific properties of the formation with respect to such other important reactions as cation exchange (Appelo & Postma 1993; Baeyens & Bradbury 1994; Pearson & Scholtis 1995). If the pore water contains solutes with concentrations that are not determined by mineral–water reactions (concentrations that are intensive properties of the system, in the thermodynamic sense, or *mobile* elements in the sense of, e.g. Michard (1987)), these concentrations or other intensive properties of the system must be known before such equilibrium calculations can be made.

Intensive properties to be evaluated could be the CO_2 partial pressure, in a system in which carbonate mineral equilibria are important, as well as the chloride and (or) sulphate concentrations of the pore water, if these substances are not determined by water–rock reactions. Measurement of *in situ* values of quantities such as $P(CO_2)$ is extremely difficult, but measurement of conservative intensive properties, such as the chloride compositions of rock samples, is relatively straightforward.

The mass of an intensive constituent *i* per mass of rock can be determined by aqueous leaching of a known mass of disaggregated sample, as is done for out-diffusion measurements. This result can be converted to the concentration in the fluid in which geochemical reactions occur using a restatement of equation (16):

$$C_{geochem}^i = C_r^i \rho \frac{1}{n_{geochem}^i} \qquad (18)$$

where $C_{geochem}^i$ is the mass of *i* per volume of geochemically reacting pore fluid, C_r^i is its mass per mass of bulk sample, determined by aqueous leaching and equation (15), ρ is the bulk density, and $n_{geochem}^i$ is the geochemical porosity for *i*.

This equation is based on the assumption that the geochemical porosity of a mudrock has the same value as the porosity that is measured by out-diffusion experiments with non-reactive dissolved substances. As discussed below, the porosity available for diffusion of solutes and for geochemical reactions may vary with the salinity of the fluid used for testing the medium. Thus, values for diffusion porosity may vary with the conditions under which they were measured. Geochemical porosity, on the other hand, refers to the volume of fluid available for geochemical reactions within the rock *in situ*. Generally, it will be the same as the diffusion porosity only when the diffusion experiments are carried out with fluid compositions like those of the *in situ* pore fluid.

Several lines of evidence indicate that the geochemical porosity of a mudrock is related to the salinity of its pore fluid. A first example can be taken from observations of the chemistry of water expelled from clay-rich material under pressure. It has been found that there is a decrease in the salinity of pore water expelled under increasing pressure (e.g. Entwisle & Reeder 1993). This decrease is attributed to the fact that water expelled first under the lowest pressures is dominantly 'free' water, whereas that expelled under higher pressures includes an increasing proportion of 'structured' (adsorbed and interlayer) water. Solutes will tend to be excluded from the structured water and so will be concentrated in the free water. Thus, the salinity of expelled water will decrease as the proportion of lower-salinity, 'structured' water increases with increasing pressure. This process was described by Appelo & Postma (1993) in terms of the double-layer behaviour of solutions at the surfaces of charged particles.

A second example can be taken from the civil engineering literature concerning the swelling of clay-rich materials when they are exposed to fluids of lower salinity than their *in situ* pore water. To explore this phenomenon, Morgenstern & Balasubramanian (1980) compared the swelling pressures developed in samples from two shale formations when they were infused with artificial natural pore waters and with distilled water. Using a model based on double-layer repulsion, they were able to account quantitatively for the higher swelling pressures they observed in the samples infused with distilled water.

To determine the geochemical porosity of a mudrock under *in situ* conditions, the concentration of the non-reactive species i in the pore water, C^i_{geochem}, must be found and combined with the concentration of i in the bulk rock using equation (18). Pore-water concentrations are best measured on field samples from boreholes in the unit of interest. Although mudrocks typically yield little if any water to boreholes, determinations of non-reactive solutes such as Cl^- are often possible, even when concentrations of other solutes have been so disturbed by contamination with drilling fluid, exposure to air, or the effects of pressure release, as to be useless for detailed characterization of the *in situ* pore water.

When no borehole samples are available, it may also be possible to determine C^i_{geochem} values on water squeezed from rock samples. This approach may be less reliable than the measurement of borehole samples because of the tendency mentioned above for squeezed water to become less saline as squeezing proceeds.

Van der Kamp *et al.* (1996) described a type of incremental out-diffusion experiment they called the radial diffusion method. They interpreted the results of their experiments in terms of pore fluid concentrations and geochemical porosity, which they referred to as effective porosity. They presented the results of experiments using HDO, chloride and sulphate, and compared them with solute concentrations from borehole data and porosities from water contents. The effective porosities for HDO are the same as the water content porosities, and the HDO concentrations from the experiments are the same as those measured in the borehole samples. This is consistent with the through-diffusion study of the London Clay, described above, in illustrating that HDO has access to the entire water content of a sample and can completely exchange with it. For chloride, the geochemical porosity for six samples is from 0.45 to 0.63 of the water-content porosity, but the chloride concentrations measured by the diffusion method are 1.15–1.55 times those measured in the borehole samples. Relative porosities and relative concentrations of sulphate are also given and range from 0.77 to 1.12 and from 0.75 to 1.15, respectively. The fact that sulphate has a higher effective porosity than chloride in spite of its higher charge and larger size seems unlikely. If slight sulphate retardation were occurring, it would account for the apparently high effective porosity (equation (10), above). The fact that the effective porosity of chloride is about half the water-content porosity is consistent with measurements on other mudrocks. It is not understood why chloride concentrations from the radial diffusion method are so much larger than those of borehole samples.

Comparison of geochemical and other porosity values

In addition to the studies described above on the London Clay by Bourke et al. (1993) and on clay-rich, glacial till by Van der Kamp et al. (1996), data on several other mudrock units can be used to compare values of geochemical and other types of porosity. Units selected as examples are the Palfris Marl in central Switzerland, the Opalinus Clay in northwestern Switzerland, and the Boom Clay in Belgium. All three units are being studied as potential host rocks for radioactive waste repositories.

Table 2 is a summary comparison of geochemical porosity values for the five example mudrocks with values for other types of porosity. The table indicates the methods by which each type of porosity was measured, and the measurement conditions. The mudrocks and porosity measurements are described in the remainder of this section.

London Clay

The diffusion measurements made by Bourke et al. (1993) are described above and their results given in Table 1. For this material, the geochemical porosities of diffusing substances are assumed to equal their diffusion porosities. The through-diffusion and out-diffusion porosities for HDO and HTO have virtually the same values as the water-content porosity, so the ratios in the last column of the table are about one. The through-diffusion and out-diffusion porosities for I^- are 0.3–0.5 of the water-content porosities. The in-diffusion results give lower ratios of HDO porosity (0.8) and of I^- porosities (0.23–0.40).

Clay-rich till

These data are results of the incremental out-diffusion (radial diffusion) study by Van der Kamp et al. (1996) discussed above. This method leads to a value for the geochemical porosity

Table 2. *Comparison of values of geochemical porosity with other measured porosity values for several mudrocks*

Material; data source	Other porosity and measurement technique	Geochemical porosity and measurement technique	Ratio of geochemical porosity to other porosity
London Clay Bourke et al. (1993)	Water content of diffusion sample c. 0.6	Through-diffusion: HDO 0.61 ± 0.05 HTO 0.59 ± 0.03 I^- 0.21 ± 0.06 Out-diffusion: HDO 0.59 ± 0.05 HTO 0.62 ± 0.06 I^- 0.32 ± 0.06 In-diffusion: HDO 0.48 I^- 0.14, 0.24	1.02 0.98 0.35 0.98 1.03 0.53 0.80 0.23, 0.40
Clay-rich till Van der Kamp et al. (1996)	Water content 0.32 ± 0.05	Incremental out-diffusion ('radial diffusion') porosity HDO 0.33 Cl^- 0.17 (calculated from porosity ratios in reference)	1.03 ± 0.04 0.53 ± 0.08 (ratios given in ref., table 5)
Boom Clay De Cannière et al. (1996)	(Not available for samples used for diffusion experiments)	Through-diffusion; Cl^- solutions of various ionic strengths HTO: 0.02 M Cl^- 0.46 ± 0.02 1.0 M Cl^- 0.59 ± 0.07 Br^-: 0.02 M Cl^- 0.18 ± 0.03 I^-: 0.02 M Cl^- 0.19 ± 0.01 1.0 M Cl^- 0.30 ± 0.09	Calculated using HTO at 0.02 M Cl^- as other porosity 1 1.26 0.40 0.43 0.70
Opalinus Clay (see text)	Water content 0.14–0.20	Cl^- leaching v. squeezing 0.08–0.11	0.54 ± 0.04
Palfris marl Baeyens & Bradbury (1994), Nagra (1997)	Hg porosimetry 0.029 ± 0.010 Water content 0.033 ± 0.012	Cl^- leaching v. borehole sample 0.009	0.32 ± 0.11 0.28 ± 0.10

which can be compared directly with the water-content porosity. For HDO, the ratios are one, whereas for Cl^-, the geochemical porosity is only about 0.5 that of the water-content porosity. This method also leads to a value of what should be the *in situ* concentrations of substances in the pore water. For HDO, these concentrations are virtually the same as those measured in samples from boreholes. For Cl^-, however, the Cl^- concentration from the diffusion measurements are from about 1.2–1.6 of those of the borehole samples. It is not clear why the diffusion-measured Cl^- concentrations are so high.

Boom Clay

The Boom Clay has been extensively studied as a potential host rock for a nuclear waste repository. The results of work on the properties of the formation relevant to transport have recently been summarized by De Cannière *et al.* (1996). Table 2 includes porosities measured by through-diffusion experiments with HTO, Br^- and I^- tracers. Two sets of measurements were made with the tracers present in NaCl solutions of two salinities. The more dilute solution had an ionic strength of 0.02 M, about that of natural pore water in the sample. The other solution was more concentrated, with an ionic strength of 1.0 M.

Because the 0.02 M solution is similar to formation water, the diffusion porosity for HTO can be taken as equal to the water-content porosity and used as the 'other porosity' in Table 2. On this basis, the relative porosities of Br^- and I^- in 0.02 M solution are virtually the same: 0.40 and 0.43, respectively. In the 1.0 M solution, the HTO porosity is greater than in the 0.02 M solution by a factor of 1.26. The I^- porosity at 1 M is 0.70 relative to the assumed water content.

The higher I^- porosity in the more saline solution is consistent with the discussion above. At higher salinity, the diffuse double-layer thickness decreases, allowing the presence of a greater volume of non-structured water capable of containing reactive solutes.

Opalinus Clay

The Opalinus Clay is under study by members of the Mont Terri Project group as an example of a type of host rock for a potential nuclear waste repository (Gautschi *et al.* 1993). The data used in this section are from as yet unpublished Mont Terri Project Technical Notes.

Data from which porosity values for the same or adjacent samples can be derived include water-content measurements, the chemistry of water squeezed from core samples, and the results of aqueous leaching (out-diffusion measurements) on a number of samples.

Water-content measurements on a number of samples by three laboratories lead to porosity values from 0.12 to 0.20 with means of 0.16 ± 0.01, 0.17 ± 0.02 and 0.15 ± 0.01. The pore-water chemistry of a number of samples was investigated by the British Geological Survey (BGS) using water samples expelled from sections of core under high pressures by the method described by Entwisle & Reeder (1993). Aqueous leaching (out-diffusion) studies were also carried out on adjacent sections of core. From these studies, the concentrations of the non-reactive solute Cl^- per mass of core leached, C_r^{Cl}, were calculated using equation (15). The Cl^- contents of the squeezed samples were taken as the Cl^- concentration in the pore fluids, C_{pf}^{Cl}. Geochemical porosities were calculated from these results using equation (16).

Figure 2 compares the geochemical and water-content porosities of several core samples of Opalinus Clay. The water-content porosities range from about 0.145 to 0.198, the geochemical porosities from about 0.087 to 0.108, and there is a weak tendency for the two sets of data to vary sympathetically ($r = 0.6$). The ratio of water content to geochemical porosity in Table 2, 0.54 ± 0.04, is the mean and standard deviation of the ratios of each of the points in Fig. 2.

Palfris Marl

The Palfris Marl is an argillite found in the Helvetic nappes in central Switzerland that is being studied by Nagra as a potential host rock for a repository for low- and intermediate-level radioactive waste. Data on the Palfris Marl have

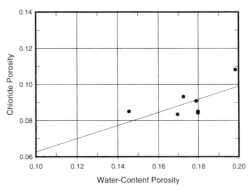

Fig. 2. Comparison of Cl^- geochemical porosity and moisture content porosity of the Opalinus Clay. The regression line through the origin is shown.

been summarized by Nagra (1997). The unit is being explored and sampled by drilling from the surface. At depths greater than 300-400 m, the formation contains zones that yield enough water that good borehole samples can be taken. At greater depths, however, the unit yields so little water to boreholes that samples, if they can be collected at all, contain large concentrations of drilling fluid, making it difficult to discern from them the properties of the formation fluid.

Geochemical modelling has been used to reconstruct the chemistry of the formation water in the zone of low water yield, on the basis of the mineralogy and cation exchange properties of core material (Baeyens and Bradbury 1994; Pearson and Scholtis 1995). The chloride content of the formation water is not controlled by water-rock reactions, so it must be considered an intensive factor in the system and specified as part of the input to the geochemical model.

Measurements of physical porosity by the mercury-injection technique, and of water-content porosity, were made on two groups of samples from Palfris core. In addition, aqueous leaching studies were carried out to measure the Cl^- content of the bulk rock. Except for a few low-porosity outliers, the mean and standard deviation of the mercury-injection values of samples of undeformed, clay-rich marl were 0.029 ± 0.010, whereas those based on water-content measurements on samples of similar mineralogy were 0.033 ± 0.012.

Aqueous leaching (out-diffusion) measurements on samples of Palfris core led to a concentration of 0.70 mM Cl^- kg^{-1} rock (Baeyens & Bradbury 1994). A pore-water Cl^- content of 0.06 M was calculated from this value using equation (18), a density of 2.65 kg l^{-1}, and an average porosity of 0.031 from the mercury-injection and water-content measurements. Samples from the zone of low water yield had Cl^- contents up to 0.16 M Cl^-, in spite of the fact that they were contaminated. This is more than twice the value calculated from the leaching data, and indicates that the porosity in which the Cl^- is accommodated in the formation, its geochemical porosity, is much smaller than the mercury-injection and water-content porosities. The extent of contamination of one sample from this zone could be determined, which fixed the Cl^- content of the pore water at about 0.2 M. From equation (16), this corresponds to a geochemical porosity of 0.009. The porosity ratios in the last column of Table 2 compare this porosity with those from the mercury-injection and water-content measurements.

Summary and conclusions

The property called porosity may have different values depending on how it is measured and the process it is meant to describe or model. Physical porosity, transport porosity and geochemical porosity can be distinguished.

Physical porosity refers to the volume of a rock that is not occupied by mineral grains. It is measured by comparing mineral grain density with rock bulk density. Total physical porosity includes isolated pores, so its value for a given rock will be the largest of any type of porosity. Values for interconnected physical porosity can be obtained by injection techniques and water-content measurements. Porosities measured by mercury injection may not include very small pores such as occur in mudrock.

Transport porosity is used when describing the movement of fluids and solutes through rock. Advection porosity relates the velocity of a fluid or perfect tracer in the pores of a rock to the fluid flux per total cross-sectional area of rock (the Darcy velocity). The advection porosity will generally be smaller than the interconnected physical porosity (water-content porosity) because the volumes of pores connected only at one end (dead-end pores) contribute to the latter but not the former.

Diffusion porosity is one of several variables used to relate the diffusion behaviour of a substance in a rock to its behaviour in pure water. Two categories of diffusion porosity can be distinguished, one related to the area across which diffusion occurs relative to the total area of rock, and the other to the volume accessible to diffusion. Area-diffusion porosity is analogous to advection porosity. Measurements of diffusion are commonly interpreted assuming that only one porosity value is needed to characterize the process.

Diffusion porosity values depend on the identity of the diffusing species as well as on the properties of the medium. Diffusion porosities for water itself, measured with the isotopic species HDO or HTO, tend to be the same as water-content porosities, indicating that all water in a rock, whether in dead-end pores or structured by inclusion in surface-bound or interlayer water layers, is available for exchange with the diffusing substance. Diffusion porosities for solutes are smaller than water-content or water-diffusion porosities, indicating that solutes do not have access to the entire water content of a rock. Inaccessible water may occur in pores into which solutes cannot pass, because of their size or ionic charge, for example. It can also be water structured by association with interlayer cations or part of the double layers formed on charged mineral surfaces. The fact that the diffusion porosity of ions increases with the ionic strength of the experimental solution supports this interpretation of the nature of inaccessible water.

Geochemical porosity is used to relate changes in solution concentrations that result from water–rock reactions and transport processes to changes in the masses of constituents per unit of rock. It is not commonly recognized as a separate type of porosity, but experimental data show that in mudrocks, at least, it is important to distinguish it from the other types. The fluid volume included in the geochemical porosity is that in which reactions occur. For mudrocks, in which solute transport can be assumed to be principally by diffusion, it seems reasonable to identify geochemical porosity with solute diffusion porosity. Diffusion porosity varies with the ionic strength of the measuring solution, however, so that if a geochemical porosity is being used to describe natural water–rock reactions, the diffusion porosity from which it is estimated should be that measured at the ionic strength of the formation fluid.

Studies of several mudrocks show that they have geochemical porosities (based on Cl^-, Br^- or I^-) that are 0.3–0.7 of other measures of porosity. For transport calculations, uncertainties or errors by factors of two or even five may be relatively unimportant when compared with uncertainties in other system properties, such as hydraulic conductivity.

Errors or uncertainties in geochemical porosity translate directly to uncertainties in the concentrations of solutions calculated from water–rock reactions. Factors of two to five in the concentrations of ions such as Cl^- that are commonly dominant in the pore waters of mudrocks, yet are not buffered by any water–rock reaction, can lead to even larger errors in the concentrations or ratios of substances such as exchangeable cations that depend on the ionic strength of the solution.

To make acceptably accurate calculations of water–rock reactions and transport in mudrock, it is important to define the geochemical porosity of the system.

The Waste Management Laboratory of the Paul Scherrer Institute develops and tests models and acquires specific data relevant to performance assessments of Swiss nuclear waste repositories. These investigations are undertaken in close co-operation and with the partial financial support of the Swiss National Co-operative for the Disposal of Radioactive Waste (Nagra). I thank my colleagues at the Waste Management Laboratory and Nagra for discussions that refined many of my thoughts on porosity. I am particularly grateful to A. Gautschi, M. Mazurek and C. Neuzil for their comments on a draft of this paper.

References

APPELO, C. A. J. & POSTMA, D. 1993. *Geochemistry, Groundwater and Pollution*. Balkema, Rotterdam.

BAEYENS, B. & BRADBURY, M. H. 1994. *Physicochemical characterisation and calculated in situ porewater chemistries for a low permeability Palfris marl sample from Wellenberg*. Nagra, Wettingen, Switzerland, Technical Report **94-22**.

BARONE, F. S., ROWE, R. K. & QUIGLEY, R. M. 1990. Laboratory determination of chloride diffusion coefficient in an intact shale. *Canadian Geotechnical Journal*, **27**, 177–184.

BOURKE, P. J., JEFFERIES, N. L., LEVER, D. A. & LINEHAM, T. R. 1993. Mass transfer mechanisms in compacted clays. *In*: MANNING, D. A. C., HALL, P. L. & HUGHES, C. R. (eds) *Geochemistry of Clay–Pore Fluid Interactions*. Chapman and Hall, London, 331–350.

BRADBURY, M. H. & GREEN, A. 1985. Measurement of important parameters determining aqueous phase diffusion rates through crystalline rock matrices. *Journal of Hydrology*, **82**, 39–55.

DE CANNIÈRE, P., MOORS, H., LOLIEVIER, P., DE PRETER, P. & PUT, M. 1996. *Laboratory and in situ migration experiments in the Boom Clay*. European Commission, Luxembourg Report, **EUR 16927**.

DOMENICO, P. A. & SCHWARTZ, F. W. 1990. *Physical and Chemical Hydrogeology*. Wiley, New York.

DORSCH, J. & KATSUBE, T. J. 1999. Porosity characteristics of Cambrian mudrocks (Oak Ridge, East Tennessee, USA) and their implications for contaminant transport. This volume.

ENTWISLE, D. C. & REEDER, S. 1993. New apparatus for pore fluid extraction from mudrocks for geochemical analysis. *In*: MANNING, D. A. C., HALL, P. L. & HUGHES, C. R. (eds) *Geochemistry of Clay–Pore Fluid Interactions*. Chapman and Hall, London, 365–388.

FREEZE, R. A. & CHERRY, J. A. 1979. *Groundwater*. Prentice–Hall, Englewood Cliffs, NJ.

GAUTSCHI, A., ROSS, C. & SCHOLTIS, A. 1993. Pore water–groundwater relationships in Jurassic shales and limestones of northern Switzerland. *In*: MANNING, D. A. C., HALL, P. L. & HUGHES, C. R. (eds) *Geochemistry of Clay–Pore Fluid Interactions*. Chapman and Hall, London, 412–422.

GELHAR, L. W. 1987. *Applications of stochastic models to solute Transport in fractured rocks*. Swedish Nuclear Fuel and Waste Management Co., Stockholm, Report **SKB TR 87-05**.

HORSEMAN, S. T., HIGGO, J. J. W., ALEXANDER, J. & HARRINGTON, J. F. 1996. *Water, gas and solute movement through argillaceous media*. OECD, Paris, Nuclear Energy Agency Report **CC-96,1**.

LEVER, D. A. 1986. *Some notes on experiments measuring diffusion of sorbed nuclides through porous media*. United Kingdom Atomic Energy Authority, Harwell, AERE Report **R 12321**.

LI, Y.-H. & GREGORY, S. 1974. Diffusion of ions in sea water and in deep-sea sediments. *Geochimica et Cosmochimica Acta*, **38**, 701–714.

MICHARD, G. 1987. Controls of the chemical composition of geothermal water. *In*: HELGESON, H. C. (ed.) *Chemical Transport in Metasomatic Processes*. Riedel, Dordrecht, 323–353.

MORGENSTERN, N. R. & BALASUBRAMANIAN, B. 1980. Effects of pore fluid on the swelling of clay–shale. *In: Proceedings of the Fourth International Conference on Expansive Soils; Characterization and Treatment of Expansive Soils for Engineering Design*. American Society of Civil Engineers, New York, **4**, 190–205.

NAGRA 1997. *Schlussbericht zu den geologischen Oberflächenuntersuchungen am Wellenberg*. Nagra, Wettingen, Switzerland, Technischer Bericht **96-01**.

NEUZIL, C. E. 1994. How permeable are clays and shales? *Water Resources Research*, **30**, 145–150.

NORTON, D. & KNAPP, R. 1977. Transport phenomena in hydrothermal systems: the nature of porosity. *American Journal of Science*, **277**, 913–936.

PEARSON, F. J. & SCHOLTIS, A. 1995. Controls on the chemistry of pore water in a marl of very low permeability. *In*: KHARAKA, Y. K. & CHUDAEV, O. V. (eds) *Proceedings of the 8th International Symposium on Water–Rock, Water–Rock Interaction — WRI-8, Vladivostok, Russia*. Balkema, Rotterdam, 35–38.

STEEFEL, C. I. & MACQUARRIE, K. T. B. 1996. Approaches to modeling of reactive transport in porous media. *Reviews in Mineralogy*, **34**, 84–129.

TRAPPENIERS, N. J., GERRITSMA, C. J. & OOSTING, P. H. 1965. The self-diffusion coefficient of water, at 25°C, by means of the spin-echo technique. *Physics Letters*, **18**, 256–257.

VAN DER KAMP, G., VAN STEMPVOORT, D. R. & WASSENAAR, L. I. 1996. The radial diffusion method. *Water Resources Research*, **32**, 1815–1822.

Permeability and fluid flow in natural mudstones

DAVID N. DEWHURST[1], YUNLAI YANG[2] & ANDREW C. APLIN[2]

[1]*T. H. Huxley School of the Environment, Earth Science and Engineering, Royal School of Mines, Imperial College of Science, Technology and Medicine, Prince Consort Road, London SW7 2BP, UK*
Present address: CSIRO Petroleum, Division of Petroleum Resources, PO Box 3000, Glen Waverley, Victoria, Australia 3150
[2]*Fossil Fuels and Environmental Geochemistry Postgraduate Institute: NRG, University of Newcastle, Newcastle upon Tyne NE1 7RU, UK*

Abstract: Mudstone permeabilities vary by ten orders of magnitude and by three orders of magnitude at a single porosity. Much of the range at a given porosity can be explained by differences in grain size; at a given effective stress, coarser-grained mudstones are more permeable than finer-grained mudstones, although the difference diminishes with increased burial. Pore size distributions illustrate why more silt-rich mudstones are more permeable than finer mudstones and also show that the loss of porosity and permeability with increasing effective stress is driven primarily by the preferential collapse of large pores. Pore size distributions can also be used to estimate permeability rapidly. None of the existing models are ideal and need to be adjusted and validated through the acquisition of a much larger permeability database of well-characterized mudstones. We also examine the role of faults and fractures as fluid conduits in mudstones. The occurrence of microscopic hydrofractures is inferred from the observation that fluid pressures in sedimentary basins rarely exceed minimum leak-off pressures. The extent to which microfractures enhance mudstone permeability, both instantaneously and over longer periods of geological time, is poorly constrained. Although fault zones in mudstones have generally low permeability, there is abundant evidence for episodic flow along faults in tectonically active regions. The role of faults as fluid conduits during periods of tectonic quiescence is less certain, and the timing and extent of any enhanced permeability and enhanced flow are not well known. In general, conditions conducive to fluid flow along muddy faults include an increase in the activity of the fault, high fluid pressures within the fault zone and the extent of overconsolidation and lithification of the mudstones.

Permeability, a measure of the ease with which fluid can flow through a porous medium, is the coefficient K (LT^{-1}) which relates the rate of fluid flow q ($L^3 T^{-1}$) to the imposed hydraulic gradient i in Darcy's law:

$$q = KiA \quad (1)$$

where A is the cross-sectional area (L^2). The limited number of published mudstone permeabilities indicates a huge variation of at least ten orders of magnitude (Neuzil 1994). However, there is very little quantitative understanding of the variation at either the microscopic and macroscopic scale, not least because there are extremely few reliable permeability data for well-characterized mudstones.

The lack of basic data is a major impediment to the accurate prediction and simulation of fluid flow and overpressure development in sedimentary basins. As a result, mudstone permeability in basin simulators is usually expressed as a simple function of porosity and it is often assumed that a single porosity–permeability function is sufficient for the very large variety of mudstone types. An indication of the current level of uncertainty is shown in Fig. 1, which displays some default porosity–permeability relationships used for 'standard shales' in some commercial basin simulators. The predicted permeabilities of the 'standard shales' vary by several orders of magnitude at low porosities.

There is also a range of opinions about the role of faults and fractures as fluid conduits within and across mudstone sequences. Although there is little doubt that both faults and fractures episodically conduct fluid through mudstones (e.g. Hubbert & Rubey 1959; Moore *et al.* 1988;

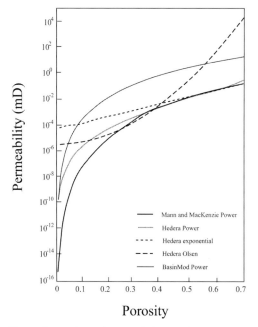

Fig. 1. Comparison of some of the porosity–permeability relationships for 'standard' shales used in computer simulators of fluid flow in sedimentary basins ('basin models'). Taken from Olsen (1962), Mann & Mackenzie (1990), Maubeuge & Lerche (1994), and Luo & Vasseur (1996).

Roberts & Nunn 1995), the extent to which faults and fractures alter both the current and long-term permeability of a mudstone sequence is poorly known, difficult to predict and exceedingly hard to model realistically.

The aim of this paper is to provide a framework for rationalizing some of the variations in mudstone permeability. Having summarized the published permeability data, we present a range of permeability models which demonstrate the central importance of pore size distribution as a control on permeability. We then show that the fundamental influences on the pore size distribution of mudstones are grain size distribution and compaction. Finally, we examine the possible role of faults and fractures as both microscale and macroscale fluid conduits in mudstones.

In this paper, we define mudstones as siliclastic sedimentary rocks comprising predominantly silt-sized (<62.5 μm) and clay-sized (<2 μm) particles. A clay is similarly defined in terms of size, but refers to a non-indurated sediment. Clay fraction and clay content are used in their strict soil mechanical meanings, i.e. the percentage of particles <2 μm (regardless of mineralogy) and percentage of clay minerals present (regardless of size), respectively. The reader should note that the majority of this paper is concerned with the water permeability of these sediments under geological stress conditions, and that mechanisms controlling permeability of mudrocks to other fluids, especially non-polar fluids (e.g. gaseous or liquid hydrocarbons), may be distinctly different.

Permeability data

Neuzil (1994) collated the few permeability data for clays and mudstones that satisfy a range of criteria related to the quality of data and their applicability to natural systems. The experimental methods used to measure permeability have been fully described in the relevant papers and as such, will not be reviewed further here. Neuzil's compilation, with additional data from Dewhurst *et al.* (1995, 1998*a*), Katsube & Williamson (1994) and Schlömer & Krooss (1997), is reproduced in Fig. 2 and suggests a log–linear relationship between porosity and permeability over a wide range of compaction states from recent marine clays to indurated mudstones. The data fall in a band approximately three orders of magnitude wide, spanning porosities between 10–80%, with hydraulic conductivities between 10^{-16} and 10^{-10} m s^{-1} at porosities between 5 and 40%. Neuzil (1994) suggested that hydraulic conductivities of around 10^{-16} m s^{-1} may be close to the minimum for mudstones.

One limitation of Neuzil's compilation is a lack of sample information, reflecting the fact that there are few permeability data for mudstones that have been characterized in terms of porosity, pore size distribution, grain size distribution and mineralogy. However, Dewhurst *et al.* (1995, 1998*a*) have recently reported data implying that much of the permeability range seen at a given porosity might be attributed to variations in the grain size distribution of mudstones. Their experimental data are shown in Fig. 3 and demonstrate that at a given porosity, samples with a lower clay fraction are about two orders of magnitude more permeable than clay-rich samples, a range that approaches that defined by Neuzil's (1994) dataset (Fig. 2). No correlation was noted with other factors such as total organic carbon content. The two groups of data fall on converging paths, as the rate of change of permeability with decreasing porosity is greater in the silt-rich samples (Fig. 3). This indicates that the permeabilities of different mudstone types converge at very low porosities, an inevitability as all muds theoretically have zero permeability at zero porosity.

The permeability data for a number of London Clay samples are plotted against effective stress in

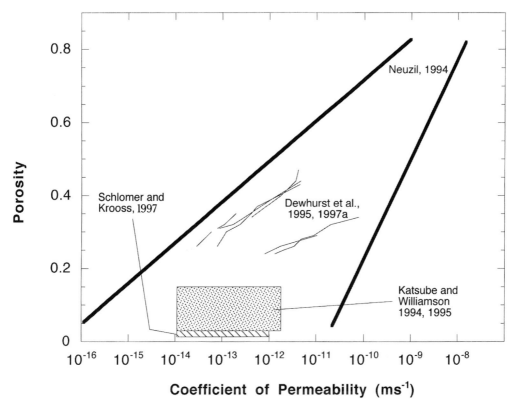

Fig. 2. Range of published porosity–permeability data for mudstones. The two bold lines define the range of values published by Neuzil (1994). Each of the fine lines denotes data published by Dewhurst et al. (1995, 1998a) for a single, experimentally compacted mudstone. Other recent data are noted in the figure.

Fig. 4, not only showing the logarithmic decrease of permeability with logarithmically increasing effective stress, but also reinforcing the important control of grain size on permeability. In this case the permeability of clay-rich samples is about 50 times lower than that of the silt-rich samples at any given effective stress.

Field-scale mudstone permeabilities, both measured and modelled, appear in some cases to be greater than those measured on core plugs in the laboratory. For example, Bredehoeft et al. (1983) documented a regional absolute permeability of 10^{-9} m s^{-1} in the Pierre Shale, although at the laboratory scale, the matrix permeability was 10^{-13} m s^{-1} (Neuzil 1993). This type of dependence has also been noted by Rudolph et al. (1991) and Keller et al. (1989) in lacustrine clays and clay-rich tills, respectively, and may be implied in other situations where abnormally large permeabilities have been reported (e.g. Davis 1988). The apparent scale dependence implies that some mudstone sequences have a heterogeneous permeability structure usually attributed to fractures, faults or the interlayering of more permeable, coarser-grained sediments. In other cases, the effects of scale appear to be small (e.g. Brace 1980; Davis 1988), implying a more homogeneous flow medium. Neuzil (1994), using data obtained from inverse analyses which estimate the properties of flow systems using numerical simulations, found that permeability was scale dependent in argillaceous media at only the largest regional scales, perhaps suggesting the existence of widely spaced transmissive fractures. We return to this issue later when discussing the role of faults and fractures as barriers or conduits to fluid flow.

Permeability models

The difficulty of routinely measuring the low permeabilities of mudstones has led to the development of a variety of models that seek to

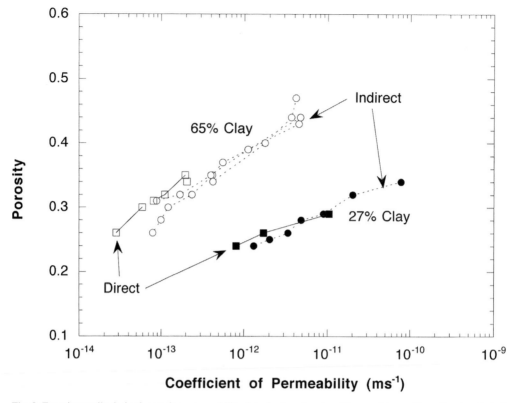

Fig. 3. Experimentally derived porosity–permeability data for two London Clay mudstones (from Dewhurst *et al.* 1998*a*). At a given porosity, the siltier (lower clay fraction) sample is much more permeable than the clay-rich sample. The permeabilities converge at lower porosities.

relate permeability to more easily measured variables such as porosity or pore size distribution. Two general approaches have been used: (1) empirical, using laboratory measurements to build relationships between permeability and porosity (e.g. Tavenas *et al.* 1983*a, b*; Nagaraj *et al.* 1990; Tokunaga *et al.* 1994); (2) theoretical, using equations based on the Kozeny–Carman and Hagen–Poiseuille equations, which describe Newtonian flow in a straight tube of circular cross-section (e.g. Leonards 1962).

Empirical models

Over recent years, many attempts have been made to correlate permeability with easily determined physical characteristics of clays. Although this has been achieved with some success in sands, using such characteristics as grain size distribution, attempts to do this with clays have met with limited success. Tavenas *et al.* (1983*a*) showed that there is no simple relationship between permeability and porosity or clay fraction. Tokunaga *et al.* (1994) proposed an equation for stresses of interest to geological conditions deep in the sediment pile, such that

$$K = K_0 \left[\frac{n}{n_0}\right]^a \qquad (2)$$

where K is current permeability, n is current porosity, K_0 is the permeability at n_0, the initial porosity, and a is a constant determined experimentally.

These relationships are generally described by a lithology-dependent variable but the approach is fundamentally limited both by the difficulty of measuring the very low permeabilities typical of mudstones and by the wide range of mudstone facies. None of these models have been verified using data for lithologically well-characterized mudstones, and the control of lithology on the variables is not known.

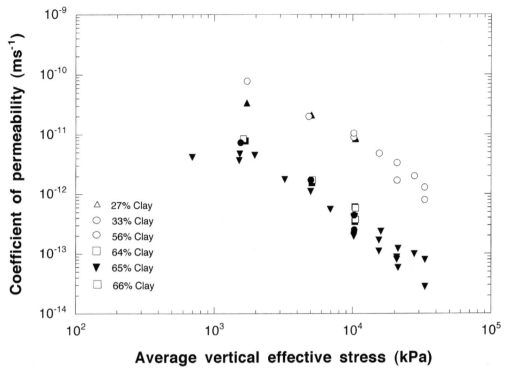

Fig. 4. Experimentally derived hydraulic conductivities of several London Clay mudstones, plotted as a function of effective stress. Although the data for each sample define a straight line, the hydraulic conductivities of the more clay-rich samples are lower at a given effective stress. Data from Dewhurst *et al.* (1995).

Theoretical model: Kozeny–Carman equation

Predictive equations remove the need to perform lengthy laboratory tests. The Kozeny–Carman equation (Carman 1956) has been used fairly successfully to predict the permeability of sands:

$$K = \frac{1}{k_0 k_T} \frac{n^3}{S^2(1-n^2)} \quad (3)$$

where k_0 and k_T are poorly defined shape and tortuosity factors, n is porosity and S is the specific surface area (in m^2 m^{-3}). However, this equation often produces discrepancies of many hundreds of per cent between predicted and measured permeability in muds (Fig. 5). This is partly because the Kozeny–Carman equation assumes that all pores are capillary tubes with the same cross-sectional area and thus that all pores have the same diameter. In reality, mudstone pores have a wide distribution of diameters (e.g. Olsen 1962; Heling 1970; Borst 1982; Griffiths & Joshi 1989, 1990; Katsube & Williamson 1994, 1995; Dewhurst *et al.* 1998*a*).

Furthermore, as the flow rate through a pore is exponentially proportional to its radius, most flow occurs through the larger, interconnected pores, which contribute only a small percentage of the surface area (Yang & Aplin 1998; Fig. 6). The total surface area is thus unlikely to be the most appropriate value for insertion in the Kozeny–Carman equation. In addition, problems arise because different adsorbants (N$_2$; ethylene glycol) give different results for specific surface area, especially if swelling clays are present (Van Olphen 1977).

Theoretical models based on pore size distributions

Several permeability models are based on the Hagen–Poiseuille equation (Leonards 1962):

$$q = \frac{\gamma I}{8\eta} ay^2 \quad (4)$$

where γ is the unit weight of the fluid, y is the radius of the tube, I is the hydraulic gradient, a is the cross-sectional area of the tube and η is the

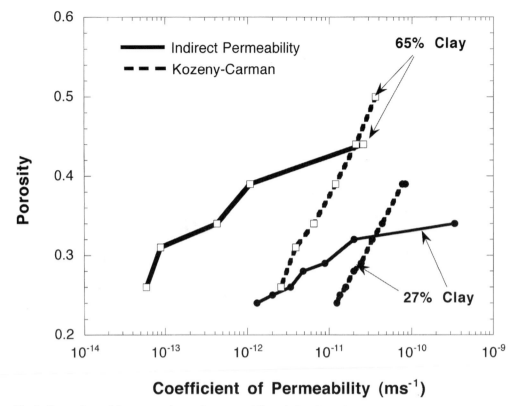

Fig. 5. Comparison of the measured porosity–permeability relationships of two London Clay mudstones (see Fig. 3) and the permeabilities calculated using the Kozeny–Carman equation.

viscosity of the fluid. This equation has been used extensively as the basis for deriving permeability from pore size distributions. Many models (e.g. Garcia-Bengoechea et al. 1979; Garcia-Bengoechea & Lovell 1981; Juang & Holtz 1986a; Lapierre et al. 1990) are based on equation (4) and are similar to that described by Scheidegger (1974):

$$K = \frac{n}{96} \int_0^\infty D^2 f(D) \, dD \quad (5)$$

where n is the porosity and D is the diameter of pores.

This model assumes that all pores are capillary tubes with circular cross-sections and that one-third of pores are aligned along the direction of flow. Four versions of the model expressed by equation (5) have been proposed: Capillary (Leonards 1962), Hydraulic Radius (Leonards 1962), Probabilistic Marshall (Marshall 1958) and Probabilistic Juang and Holtz (Juang & Holtz 1986a). All describe flow through a porous medium with a range of pore diameters, d_i, whose volumetric frequency, $f(d_i)$, is such that

$$\sum_i f(d_i) = 1$$

where $f(d_i)$ can be obtained directly from mercury intrusion tests.

Both Garcia-Bengoechea et al. (1979) and Juang & Holtz (1986b) found that these predictive equations gave good results for the permeability of sand–clay mixtures. In contrast, Lapierre et al. (1990) found that none of the models accurately predicted the permeability of natural clays, often overestimating permeabilities by up to two orders of magnitude or more (Lapierre et al. 1990). There are several possible reasons for the observed discrepancy. First, the models are derived from isotropically permeable, mechanically compacted soils, whereas naturally compacted clays are anisotropic (Leroueil et al. 1990). Second, the mathematical descriptions assume that pores are continuous tubes, whereas in practice pore size distributions measured by

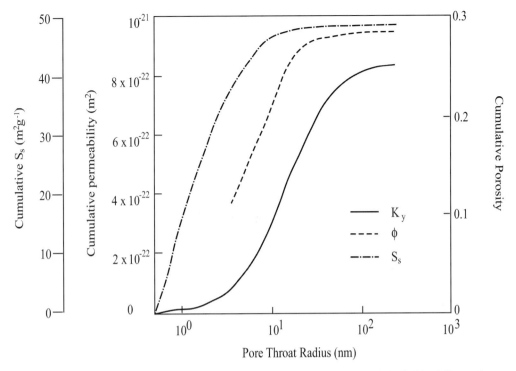

Fig. 6. Cumulative vertical permeability (K_v), porosity (ϕ) and specific surface area (S_s) of a North Sea mudstone, calculated using the pore model of Yang & Aplin (1998). The data show that most of the surface area of the mudstone is in the smallest pores, whereas most of the permeability results from the largest pores.

mercury intrusion determine the size distribution of pore throats. The pore throat is smaller than the pore and therefore in a real pore system more than one model pore tube is required to account for the volume of the real pore. As a result the modelled permeability will be overestimated.

Yang & Aplin (1998) have addressed these difficulties by developing a permeability model that is based on the Hagen–Poisueille equation but defines a pore shape as two frustrums of cones, connected at their base. The pore shape is allowed to change such that its aspect ratio increases as a function of effective stress. The variables defining the pore shape are thus dependent on effective stress and lithology. Permeability anisotropy is allowed to increase with increasing effective stress, most particularly for clay-rich samples with high values of the compressive coefficient β. This is consistent with experimental observations of increasing permeability anisotropy with increasing effective stress and decreasing silt content in consolidated clays (Clennell et al. this volume). The absolute levels of anisotropy used in the model are similar to those measured in laboratory tests of consolidated clays, which are around unity for silty clays, 2–3 in kaolinite, 4–8 in smectite and usually less than four in natural clays (Freeze & Cherry 1978; Garcia-Bengoechea et al. 1979; Garcia-Bengoechea & Lovell 1981; Tavenas et al. 1983a, b; Al-Tabbaa & Wood 1987; Znidarcic & Aiban 1988; Lapierre et al. 1990; Leroueil et al. 1990; Clennell et al. this volume).

All the models predict different permeabilities and we are currently in the unhappy situation of not knowing which model gives the most accurate predictions and under what range of conditions. This situation will remain until more permeability measurements are made on carefully characterized mudstones.

One of the useful facets of permeability models based on pore size distributions is that they allow the contributions of individual pores to total fluid flow to be calculated. This is illustrated in Fig. 6, which shows plots of cumulative permeability and cumulative surface area as functions of porosity, calculated using the pore shape model of Yang & Aplin (1998). Most of the permeability resides in the largest pores (>100 nm), whereas most of the surface area lies in very small pores with diameters <10 nm. There are two important implications of these data. First, they force one to question the applicability of the Kozeny–Carman equation to the prediction of mudstone permeability, as neither the total porosity nor the specific surface area is directly related to the fluid transport properties of mudstones. Second, they lend confidence to the idea that Darcy's law is likely to accurately describe fluid flow in mudstones. Although Darcy's law has been shown to be appropriate in laboratory tests, its applicability has not been fully tested at the lower hydraulic gradients observed in nature. The applicability of Darcy's law has also been questioned because of the possibility that a significant fraction of water in highly compacted mudstones may be bound to mineral surfaces and thus exhibit different physical properties from those of free water. Although there is still debate about the distance to which mineral surfaces exert an influence on the nature of pore water (e.g. Drost-Hansen 1991; Israelachvili 1992), typically quoted values are around 2–3 nm (Newman 1987). As pores of this size contribute little to the overall permeability of mudstones (Fig. 6; Yang & Aplin 1998), it is likely that Darcy's law reasonably describes fluid flow in natural mudstones, just as it does in laboratory experiments on natural mudstones (e.g. Olsen 1965; Neuzil 1986, 1994; Dewhurst et al. 1998a).

Pore size distributions of natural mudstones

By showing that flow rate increases rapidly with pore radius, the Hagen–Poiseuille equation illustrates the predominant control that pore size distributions exert on mudstone permeability. The purpose of this section is thus to explore the main controls on the pore size distributions of mudstones and how they change with increasing compaction.

Our basic hypothesis is that the pore size distributions of mudstones are fundamentally controlled by two variables: porosity (extent of compaction) and grain size distribution. As permeability is strongly dependent on pore size distribution, this statement also implies that permeability is influenced mainly by the same two variables. Other researchers have stressed the influence on permeability of clay microstructure (e.g. Mitchell et al. 1965; Neuzil 1994), aggregated structures (Delage et al. 1982; Delage & Lefebvre 1984; Tavenas et al. 1983a; Lapierre et al. 1990; Leroueil et al. 1990) and the interlayer cation in smectite (Lambe & Whitman 1979; Di Maio 1996). These factors may be relevant in shallow-buried muds and where bentonite is being used to contain waste. However, any influence that microstructure and aggregation exert is likely to decline with increasing stress (Griffiths & Joshi 1990). Furthermore, the quantitative assessment of clay microstructure is difficult and begs the question of what variables define it. We feel that grain size and grain shape are key factors in defining fabric and should therefore be treated as the primary controls on pore size distribution and thus fluid flow.

Most published pore size distributions have been determined using mercury-intrusion porosimetry, which measures the diameter of pore throats rather than that of pore bodies. Although unsatisfactory for some purposes, mercury injection is an appropriate measure of pore size for the purpose of estimating permeability, as it is the pore throats that control fluid flow in porous media.

Data from both geotechnical and geological literature show that the mean pore throat diameter of muds declines with increasing depth or effective stress, typically reaching values of <10 nm at depths of 3–5 km (Heling 1970; Borst 1982; Griffiths & Joshi 1989, 1990; Katsube & Williamson 1994, 1995; Dewhurst et al. 1998a). The shift to a smaller mean pore size is driven primarily by collapse of large pores, with smaller pores remaining relatively unchanged (Delage & Lefebvre 1984; Griffiths & Joshi 1989, 1990; Lapierre et al. 1990; Dewhurst et al. 1998a). This effect is shown clearly by the results of experiments in which two mudstones from the London Clay with distinct grain size distributions were experimentally compacted (Fig. 7a; Dewhurst et al. 1998a). Although both samples show how large pores collapse preferentially, the effect is more obvious in the coarser-grained sample (Fig. 7b) because in its initial state (preconsolidation stress of around 1.5 MPa) it has both a broader pore size distribution and larger pores.

A major problem in interpreting the published pore-size data for mudstones is that few of the publications give sample descriptions; all samples are simply 'mudstones' or 'shales'. Some of our recent work has looked at the effect of the grain size of mudstones on pore size distributions, both for natural samples and for samples compacted in the laboratory. Measurements on both experimentally and naturally compacted mudstones show the strong influence of lithology on pore size distributions. The London Clay data in Fig. 7 show that at a given effective stress, the mean and modal pore sizes are much greater in the coarser-grained sample than in the finer-grained sample. The clay-rich sample contains almost no pores with radii >100 nm, even at stresses as low as

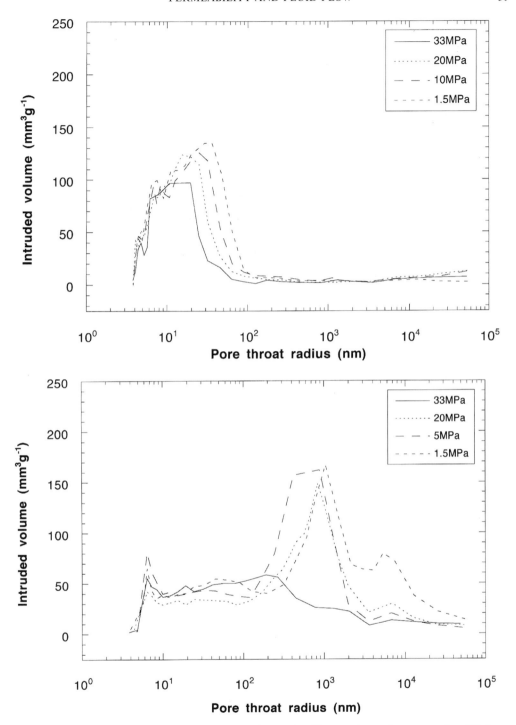

Fig. 7. Pore size distributions of two mudstones from the London Clay experimentally compacted to differing levels of effective stress. The sample in (**a**) has a clay fraction of 65%. There are few pores with radii greater than 100 nm and larger pores are lost at increasing effective stresses. The sample in (**b**) has a clay fraction of 27%, and has both a proportion of much larger pores and a much broader pore size distribution than the finer-grained sample. Larger pores collapse with increasing effective stress. Data from Dewhurst *et al.* (1998*a*).

1.5 MPa. These data imply that larger pores are more stable and more difficult to collapse in coarser-grained mudstones than in their finer-grained counterparts.

The differences in pore size distributions account qualitatively for the permeability difference between the two sediments (Fig. 3), and the more obvious and greater change in the pore size distribution of the siltier sample also explains the more rapid decline in its permeability during compaction.

Pore-size data for naturally compacted North Sea mudstones (Yang & Aplin 1998) support the conclusions of Dewhurst *et al.* (1998a) that lithology and porosity jointly exert major controls on the pore throat radii of mudstones. Figure 8 shows pore size distributions of four lithologically similar (50-55% clay fraction) mudstones that have been buried to different effective stresses. Both the porosity and the median pore radius decrease with increasing stress, with the median pore radius decreasing from 18 nm at 15.9 MPa effective stress to 6 nm at 27.5 MPa (Fig. 8). In contrast, Fig. 9 shows the pore size distributions of three lithologically distinct mudstones that have all been buried to the same level of effective stress (7 MPa). Total porosities range from 48% for a silt-rich sample to 29% for two clay-rich samples, and the median pore size ranges from 80 nm (silt-rich sample) to 5 nm (most clay-rich sample). The siltier mudstone has a higher porosity and larger pore throats than the clay-rich samples.

The effect of the different pore size distributions on the calculated permeabilities of these naturally compacted mudstones is shown in Figs 10 and 11. The calculated permeabilities of the four lithologically similar mudstones (whose pore size distributions are illustrated in Fig. 8) decreases logarithmically with increasing effective stress and porosity (Fig. 10). The calculated permeabilities of the three lithologically distinct samples vary at the same effective stress by more than two orders of magnitude, as a direct func-

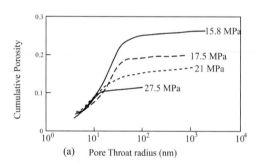

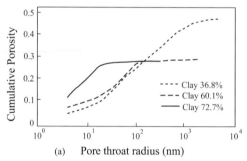

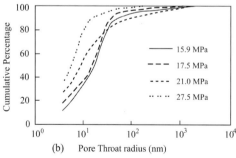

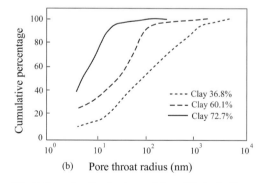

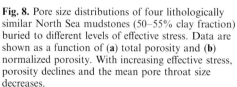

Fig. 8. Pore size distributions of four lithologically similar North Sea mudstones (50-55% clay fraction) buried to different levels of effective stress. Data are shown as a function of (**a**) total porosity and (**b**) normalized porosity. With increasing effective stress, porosity declines and the mean pore throat size decreases.

Fig. 9. Pore size distributions of three lithologically distinct North sea mudstones, all buried to approximately the same effective stress (7 MPa). Data are shown as a function of (**a**) total porosity and (**b**) normalized porosity. The more fine-grained mudstones have lower porosities and a smaller mean pore throat radius.

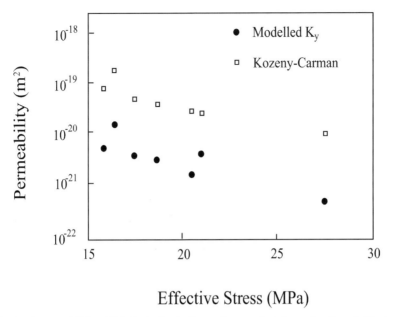

Fig. 10. Calculated permeabilities of lithologically similar mudstones plotted as a function of effective stress. Two permeabilities are shown: 'modelled K_y' is the vertical permeability calculated by the model of Yang & Aplin (1998). 'Kozeny–Carman' is the permeability calculated by the Kozeny–Carman equation. In both cases, permeabilities decrease logarithmically with effective stress but the Kozeny–Carman permeability is about an order of magnitude greater.

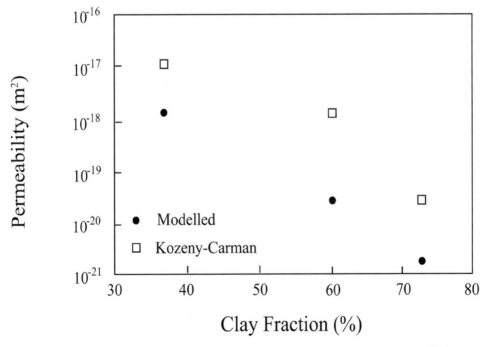

Fig. 11. Calculated permeabilities of three lithologically different mudstones buried to similar effective stresses (see Fig. 9 for pore size distributions). Two permeabilities are shown: 'modelled K_y' is the vertical permeability calculated by the model of Yang & Aplin (1998). 'Kozeny–Carman' is the permeability calculated by the Kozeny–Carman equation.

tion of their clay fractions (Fig. 11). It should be noted also that the permeabilities calculated using Yang & Aplin's model are more than an order of magnitude lower than those calculated using the Kozeny–Carman equation. At porosities below 35–45%, the Kozeny–Carman equation also predicted a much higher than measured permeability for the two samples of London Clay shown in Fig. 5.

Mudstones exhibit a much wider grain size distribution than sandstones, with grain diameters typically ranging over five orders of magnitude (e.g. Aplin *et al.* 1995; Dewhurst *et al.* 1998*a*). The broad grain size distribution is mirrored by a relatively wide spectrum of pore sizes, especially at lower levels of effective stress and in more silty muds. As fluid movement through interconnected large pores is greater than that through small pores, much of the flow in mudstones occurs through such larger, connected pores. For the samples discussed above, 90% of the permeability of the four lithologically similar mudstones buried to effective stresses above 16 MPa was calculated to be held within 50–70% of the largest pores. For the three lithologically distinct samples (36–72% clay fraction) buried to 7 MPa effective stress, 90% of the permeability is calculated to be in 30–60% of the largest pores (Yang & Aplin 1998).

Chemical diagenesis

The ideas and models presented above assume that compaction and pore collapse is driven uniquely by mechanical processes. This is a simplification, as a range of chemical processes also affects fine-grained clastic and carbonate sediments. Although the dominant clay mineral reactions (e.g. smectite to illite transition) mainly occur in the deeper parts of sedimentary basins, reactions involving carbonates occur at all stages of diagenesis. These include the formation of hardgrounds at the sea bed, the precipitation of laterally extensive carbonate bands close to the sediment–water interface (e.g. Irwin *et al.* 1977) and the chemical compaction of chalk at depths greater than *c*. 1 km (Schneider *et al.* 1993). Reactions involving carbonates generally involve the precipitation of pore-occluding cements and are likely to reduce permeability, perhaps at an early stage of burial. There is relatively little published work on this problem but some researchers (e.g. Hunt 1990; Ward *et al.* 1994) have suggested that carbonate-cemented horizons may form laterally extensive pressure seals in the North Sea.

The effect on permeability of the conversion of smectite to illite is unknown. Although in sandstones the occurrence of illite severely reduces permeability, there is no reason to believe that this will also be the case in mudstones. Furthermore, the smectite to illite transformation may lead to a slight decrease in mineral volume and results in grain coarsening (Ahn & Peacor 1986); together these factors might increase rather than decrease permeability.

These considerations show mainly how little is known about the effects of chemical reactions on mudstone fabric, porosity and pore size distribution. More detailed studies on well-characterized materials are needed to resolve the issues.

Faults and fractures

The role of faults and fractures as fluid conduits through mudstones has received extensive scrutiny from both geologists and engineers. Field studies include observations in sedimentary basins (e.g. Smith 1980; Harding & Tuminas 1988, 1989; Hooper 1991; Bjørlykke & Høeg 1997), accretionary complexes (e.g. Moore *et al.* 1990; Hill *et al.* 1993; Carson *et al.* 1995; Lewis *et al.* 1995; and references within these volumes) and landslides (e.g. Skempton 1964, 1985; Bromhead 1992). Experimental evidence has been presented by, for example, Morrow *et al.* (1981, 1984), Arch & Maltman (1990, 1993), Stephenson *et al.* (1994), Dewhurst *et al.* (1996*a, b*, 1998*b*), Bolton & Maltman (1997) and Clennell *et al.* (1998).

Paradoxically, it appears that faults in mudrich sediments can act as both efficient barriers to fluid flow (e.g. Smith 1980; Morrow *et al.* 1981, 1984; Knipe 1993, 1997; Dewhurst *et al.* 1996*a, b*; Clennell *et al.* 1998) as well as conduits for channelized migration of pore fluids (e.g. Moore *et al.* 1988; Moore 1989; Screaton *et al.* 1990; Brown *et al.* 1994; Clennell *et al.* 1998). These observations suggest that faults *intermittently* transmit fluid. The challenges are thus (a) to understand and predict the conditions under which faults are open or closed to fluid flow and (b) to estimate the rate of fluid flow and the fluid flux. In general, the permeability of muddy fault zones appears to be dependent on the close relationships between macro- and micro-fabric, pore fluid pressure, fluid flow and subsequent mechanical and chemical changes induced by compaction, progressive shear and diagenesis (e.g. Morgenstern & Tchalenko 1967; Maltman 1988; Logan 1992; Knipe 1993; Stephenson *et al.* 1994; Dewhurst *et al.* 1996*a, b*, 1998*b*; Clennell *et al.* 1998). Similar considerations apply to microfractured mudstones. What conditions promote microfracturing? What is the instantaneous increase in permeability and for how long are the high permeabilities maintained?

What volume of fluid might be mobilized as a result of microfracturing?

Microfractures

Microfractures are thought to form in mudstones when fluid pressures exceed the minimum principal (usually horizontal) stress and the tensile strength of the rock *in situ*. Microfractures have often been invoked as a way of enhancing mudstone permeability along fault zones (e.g. Sibson 1981; Behrmann 1991) and are also discussed as a way in which the general permeability of mudstones can be increased away from fault zones, potentially draining petroleum reservoirs and releasing high fluid pressures. However, a thorough literature review revealed only three directly relevant publications on the subject (Capuano 1993; Dick *et al.* 1994; Grimm & Orange 1997). Of these, only Capuano's paper deals with deeply buried mudstones and is the only one in which images of microfractures, in this case cemented with organic matter and anhydrite, are shown. Even in this case, there is some uncertainty as to the mechanisms by which the fractures formed (Milliken & Land 1994; Capuano 1994).

Evidence for microfractures in mudstones is thus primarily inferential, hinging on the observation that fluid pressures in potential petroleum reservoirs rarely exceed the formation strength as defined by leak-off tests (Gaarenstroom *et al.* 1993). During a leak-off test, fluid is pumped into non-reservoir units at a known rate and a plot of fluid pressure v. volume is constructed. Deviation from a straight-line relationship defines the leak-off pressure (LOP) at which the formation starts to fracture. Gaarenstroom *et al.* (1994) defined a minimum LOP trend for the Central North Sea and showed that there are few fluid pressures higher than the LOP, implying pressure loss by fluid flow through fractures. Gaarenstroom *et al.* also suggested a much increased risk of petroleum leakage from reservoirs that have fluid pressures within *c.* 7 MPa of the minimum LOP.

Support for the conclusions of Gaarenstroom *et al.* (1993) comes from Darby *et al.* (1996), who showed fluid pressure data from across the North Sea's UK Central Graben. In Fig. 12, minimum LOP data from Gaarenstroom *et al.* (1993) are superimposed on Darby *et al.*'s (1996) data, showing that the fluid pressures approach but never cross the LOP curve. The origin of the required high fluid pressures has been debated extensively but the most likely mechanisms are (1) pressure transfer from deeper parts of basins along connected aquifers and (2) gas generation at temperatures over *c.* 150°C (see review by Osborne & Swarbrick (1997)). Disequilibrium compaction, although contributing to overpressure, is unlikely to generate pressures close to the LOP, because by this mechanism the rate of increase of pressure as a function of depth cannot exceed the lithostatic gradient, as required if fluid pressure is to achieve the LOP.

If microfractures contribute to fluid flow in and across mudstones, important questions include the extent to which permeability is instantaneously enhanced and to what extent the increased permeabilities can be maintained through geological time. The second problem rests on the extent to which fractures remain open after formation. Both questions are difficult to answer. As the distribution of microfractures in mudstones is poorly constrained, there are few data with which to calculate permeability. Capuano (1993) used the distribution of suspected microfractures in some US Gulf Coast mudstones to calculate a fracture permeability of 10^{-13} m^2, implying a similar permeability to local sandstones. In this specific case, anhydrite cements are likely to have reduced the apparently high, instantaneous permeabilities.

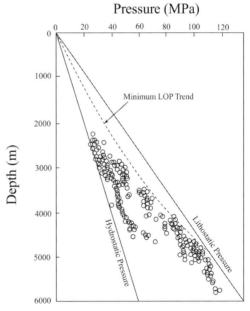

Fig. 12. Fluid pressure data for Central North Sea reservoir units (Darby *et al.* 1996). Also shown are the lithostatic and hydrostatic pressure–depth trends plus the minimum leak-off pressure (LOP) trend of Gaarenstroom *et al.* (1993). Essentially none of the fluid pressures exceed that defined by the minimum LOP, suggesting that fluid flow and pressure release occurs by hydrofracturing at these pressures.

More generally, such high permeabilities cannot be sustained over large volumes of mudstone and over significant periods of geological time because they would rapidly lead to depletion of the high fluid pressures that characterize the deeper parts of most sedimentary basins.

Appraisal of relevant Mohr diagrams (e.g. Ingram & Urai this volume) suggests that mudstones in sedimentary basins are more likely to hydrofracture by shear than by extension, and theory suggests that fractures are more likely to form horizontally rather than vertically (e.g. Lorenz et al. 1991; Brown 1994). During shear, enhanced fracture permeabilities result when mudstones dilate, rather than compact. Conditions promoting dilatant behaviour are high fluid pressures, overconsolidation and increasing rock strength, a condition that is poorly constrained in detail but that one would anticipate would be promoted by lithification resulting from chemical diagenetic changes including carbonate cementation (e.g. Irwin et al. 1977) and clay mineral recrystallization (e.g. Hower et al. 1976; Ahn & Peacor 1986). Although an understanding of these conditions is useful as a predictive tool for seal failure in petroleum systems (Ingram & Urai this volume), it is still difficult to make a quantitative assessment of both the instantaneous and long-term permeability of sheared, hydrofractured mudstones.

Faults in mudstones as seals

Reduced permeability in muddy fault zones is predicted by critical state theory (e.g. Schofield & Wroth 1968; Atkinson & Bransby 1978; Jones & Addis 1985), as the porosity of a sheared clay will be lower than its consolidated counterpart as a result of shear-induced consolidation (e.g. Arch & Maltman 1990; Clennell et al. 1998). A number of experimental studies have noted that matrix permeability across muddy fault zones is severely reduced, reaching values as low as 0.3 nD at confining pressures of up to 200 MPa (e.g. Chu et al. 1981; Morrow et al. 1981, 1984; Brown & Moore 1993). The permeability of some of these clay-rich gouges was noted to be influenced by grain size distribution, confining pressure and differential stress, but was not controlled by the finite strain accumulated during these experiments. Increasing clay contents tend to influence the mechanical behaviour of muddy faults, resulting in predominantly sliding shear behaviour (Lupini et al. 1981), which serves to further reduce permeability (Morrow et al. 1984). Further studies have noted that the matrix permeability across a muddy fault zone can be distinctly lower than that of the surrounding consolidating wall rocks by factors of up to 50, dependent on the grain size and mineralogy of the faulted material; in contrast, the matrix permeability along such a fault zone was noted to be similar to that of the wall rocks (Brown et al. 1994; Dewhurst et al. 1996a, b, 1998b). The shear-normal permeability reduction was attributed to strain localization within the fault zones, as well as increased particle alignment, both of which may result in the progressive preferential collapse of large pores.

Faults in mudstones as fluid conduits

The evidence presented above indicates that muddy faults possess intrinsically low permeability. However, numerous accounts have been documented in foreland, basinal and accretionary wedge environments which illustrate considerable degrees of fluid migration (tens of kilometres) along fault zones in clay-rich sediments (e.g. Hubbert & Rubey 1959; Moore et al. 1988, 1995; Moore 1989; Fisher & Hounslow 1990; Screaton et al. 1990; Logan 1992; Brown et al. 1994; Shipley et al. 1994; Roberts & Nunn 1995). Therefore, even though theoretical and experimental research suggests that the permeability of a fault zone in clay-rich sediments should be lower than that of the surrounding wall rocks, it appears that extensive fluid flow may occur as a result of dilation and/or fracturing of the fault zone sediments.

As compaction and diagenesis progress, clay-rich sediments become increasingly consolidated and lithified, resulting in increased brittleness with depth, which renders them more susceptible to dilation and failure (Jones & Addis 1985; Jones 1994; Ingram & Urai this volume). For an unlithified, clay-rich sediment, the level of overconsolidation controls the degree of dilation after the onset of yielding and failure (Brown 1994; Jones 1994). Normally consolidated and slightly overconsolidated clays rarely dilate, usually reducing permeability through porosity collapse, whereas heavily overconsolidated sediments dilate the most (Schofield & Wroth 1968; Jones & Addis 1985). Variations in pore fluid pressure and thus effective stress occur as a result of undrained failure of low-permeability, unlithified clay-rich sediments. As such, the low-permeability of muds governs the change in pore fluid pressure; fluid pressures will be elevated during the failure of normally consolidated clays and reduced in overconsolidated clays (Brown 1994). Lithified mudstones dilate because of the propagation of compression-induced extensional microfractures along grain boundaries and from pores and other heterogeneities within the sediment (e.g. Brace et al. 1966; Brown 1994).

Experimental evidence also suggests that faults and shear zones can be sites of preferred fluid flow (e.g. Knipe 1986; Maltman 1988). Evidence of dilation of clays and enhanced fluid flow along shear zones was documented by Arch & Maltman (1990, 1993) and ascribed to decreased tortuosity parallel to the shear zone resulting from sub-parallel particle alignment. Stephenson et al. (1994) and Bolton & Maltman (1997) both documented transient pulses of fluid expulsion during active deformation of overconsolidated clays. They inferred a component of dynamic permeability to account for the observations, driven in part by the mechanical rearrangement of the fabric during shearing, leading to pore collapse and fluid expulsion.

Fluid flow in clay-rich fault zones: field evidence

Sedimentary basins. Experimental evidence for enhanced fluid flow along muddy fault systems is supported by observations of ancient fault zones and also by Ocean Drilling Program research into currently active faults in accretionary complexes. In the Keuper Marl at Watchet in southwest England, intense veining is noted in the hanging walls and along the planes of some faults, but no veining is evident in the footwalls (Logan 1992). This implies that the faults acted as barriers to cross-fault fluid migration but as a conduit along the fault zone itself (Logan 1992). Currently inactive growth faults in compacting Gulf Coast sediments inhibit fluid flow across fault planes. However, the fault planes themselves are, in some cases, extensively mineralized, implying substantial fluid flow in the past (Hooper 1991). This episodicity of fault zone permeability indicates that growth faults appear to concentrate fluid flow during periods of fault activity, which correlate with high sedimentation rates, but that flow is impeded when such faults become inactive. Where these faults cut oil-bearing sequences, they form poor flow channels for oil when water-wet, but if the system is or becomes oil-wet, fault zone permeability to hydrocarbons could increase dramatically (Hooper 1991).

Another potential route for transmitting fluid both out of and through low-permeability sediments are the extensive ($>10^5$ km^2) intraformational polygonal fault systems which were first discovered in smectite-rich, Tertiary mudstones in the North Sea (Cartwright 1994) and have now been documented in mudstones from many sedimentary basins (Cartwright & Dewhurst 1998). Although it is still uncertain how much fluid has been transmitted through these specific fault systems (Cartwright 1994; Bjørlykke & Høeg 1997), studies of similar faults in clay-pits in Belgium have discovered Cretaceous microfossils within Tertiary clays, displaced >80 m above their stratigraphic level along faults with only 6 m throw (Verschuren 1992). Considerable vertical fluid migration along the fault planes can therefore be inferred. Furthermore, where polygonal fault systems have intersected overpressured sand bodies, fluidization has resulted in injection of remobilized sand along fault planes forming clastic intrusions, some of which are saturated with hydrocarbons (Rye-Larsen 1994).

Overall, the evidence from sedimentary basins implies periodic fluid flow along faults. By analogy with flow associated with earthquakes, enhanced fluid flow is likely when the faults are active. Enhanced fluid flow also depends on the extent to which the mudstones dilate during faulting, which is in turn influenced by their state of consolidation–lithification. Predicting the likelihood of fault-related flow in the geological past thus requires insights into the mechanical properties of mudstones and also the magnitude and frequency of fault movements at different stages of basin evolution.

Accretionary complexes. Currently active accretionary complexes are ideal settings to study fluid migration along muddy fault zones (e.g. Moore et al. 1990). The low average vertical permeabilities in the sediment wedges tend to promote extensive lateral fluid migration along stratigraphically controlled tectonic conduits. In the Barbados accretionary complex, geochemical and fluid pressure anomalies point to the role of the décollement as a barrier to vertical fluid migration in the wedge (Moore et al. 1988; Gieskes et al. 1990). Here, high thermogenic methane contents are noted in and beneath the décollement, whereas thrust faults above the décollement that are conducting fluids have no thermogenic methane signature (Gieskes et al. 1990). The underthrust sediments beneath the décollement are highly overpressured but almost undeformed, indicative of the vertical barrier to fluid migration caused by the localization of the décollement in a 40 m thick smectite-rich clay unit (Moore et al. 1988), generally the most impermeable of clay minerals (e.g. Mesri & Olson 1971; Mitchell 1993; Clennell et al. this volume).

Anomalously fresh water and the high levels of thermogenic methane detected in both the décollement proper and in the incipient décollement, seaward of the deformation front, were attributed to smectite dewatering and organic maturation, implying lateral transport of 50–70 km along the décollement, from depths (3.5–5 km) where the smectite dehydration–organic maturation window was attained (Gieskes et al. 1990; Brown 1994). Furthermore, high temperatures in some of the thrust faults and the décollement indicated movement of warm fluids along these fault zones (Fisher & Hounslow 1990). The current temperature and geochemical anomalies associated with the faults are considered to be episodic, as calculations indicate that they would die out within 10^4–10^5 years (Fisher & Hounslow 1990; Gieskes et al. 1990). The main problem associated with the extensive fluid flux along the décollement of the Barbados accretionary complex is that it comprises 80–90% smectite, which has extremely low permeability even at low effective stress and high porosity (e.g. Clennell et al. this volume). Fluid flow models of the area (Screaton et al. 1990) indicate that the permeability of the décollement must have been 3–5 orders of magnitude greater than that of the surrounding sediments to account for the observed fluid flux. The sediment in the décollement shows a scaly fabric consisting of centimetre-scale, polished and anastomosing shear surfaces, trending subparallel to the shear-zone walls. Adapting the models of Sibson et al. (1975), Sibson (1981) and Vrolijk (1987), it has been suggested that the low permeability of the sediments in the décollement results in rising pore fluid pressure leading to dilation of the scaly fabric and resultant weakening of the fault zone. This leads to instantaneous fluid movement up the fault zone and simultaneous pressure release. The low sediment permeability results again in rising pore fluid pressures and so the deformational pumping cycle continues (Moore 1989), emphasizing the coupling between tectonic and hydrogeological processes in accretionary complexes. The low angle of the maximum principal stress in the wedge also inhibits the development of vertical hydrofractures, thus preventing simple venting up through the prism to the sea bed, but allowing sub-horizontal hydrofractures to form parallel to the décollement (Moore 1989). Particle alignment and the sub-horizontal network of hydrofractures prevent fluid migration across the décollement and also help to retain laterally migrating fluids within it. However, if the whole décollement was at lithostatic fluid pressure, this would result in the collapse of the wedge. This suggests that there are areas within the décollement close to lithostatic fluid pressures, whereas other parts have substantially lower fluid pressure (Brown et al. 1994); this inference was indeed borne out by 3D seismic imaging of and subsequent fluid pressure calculations for the Barbados décollement zone (Shipley et al. 1994).

In the case of both lithified and unlithified clay-rich sediments, the yielding and failure described above may lead to the development of an anastomosing network of hydrofractures parallel to the fault zone. The development of an interconnected hydrofracture network is controlled by the mean effective stress and type of faulting (Sibson 1981; Behrmann 1991). For example, theoretical predictions suggest that thrust faulting requires much higher fluid pressures than normal or wrench faulting in accretionary wedges, and that hydrofracturing occurs concomitant with shear failure in normal and wrench faults at shallow depths in this setting (Behrmann 1991). This implies that such modes of faulting will form effective conduits for migrating fluids resulting in sea-floor venting, evidence for which is provided by the existence of chemosynthetic biological communities and mineral crusts at the outcrops of faults imaged on 2D seismic sections (Carson et al. 1991; Le Pichon et al. 1992). These observations of leaky faults in shallow unlithified sediments are further supported by geochemical evidence from the southern Nares abyssal plain (Buckley & Grant 1985; Buckley 1989a, b). Geochemical testing of cores through faults in the upper 25 m of sediment indicated extensive leaching of iron and manganese along some fault planes as compared with the surrounding wall rocks, implying the movement of reducing fluids along the faults. The observation that not all faults had leaching haloes was considered to provide evidence for episodic fluid flow (Buckley 1989a, b).

Summary

Limited laboratory and field data indicate that mudstone permeabilities vary by ten orders of magnitude, spanning a range of three orders of magnitude at a given porosity. It is likely that much of the range seen at a given porosity is related to the grain size distribution of the mudstones. At a given porosity silt-rich, coarser-grained mudstones are more permeable than finer-grained mudstones. With increasing effective stress, permeability is lost by the progressive collapse of large pores. Mudstone permeabilities probably converge at high effective stresses as a result of the homogenization of pore size distributions through preferential collapse of large pores.

As the generally low permeabilities of mudstones are difficult and time-consuming to measure, permeabilities are often estimated using a variety of empirical and theoretical models. None are ideal and some undoubtedly give poor predictions. Empirical models are based on laboratory data and have limited, general use. Many theoretical models are based on the Hagen–Poiseuille equation and use pore size distributions as their basic input. The general applicability of these models will only be established when they are more rigorously tested against data for natural samples. Similarly, the Kozeny–Carman equation must be used with caution when evaluating mudstone permeability.

Hydrofracturing of mudstones will occur when fluid pressures exceed the minimum principal stress and the tensile strength of the rock. Although microfractures have rarely been observed in deeply buried mudstones, their occurrence is inferred from the observation that fluid pressures in sedimentary basins rarely exceed minimum leak-off pressures. The extent to which microfractures enhance mudstone permeability, both instantaneously and over longer periods of geological time, is poorly constrained. Enhanced fracture permeabilities result when mudstones dilate, most likely under conditions of high fluid pressures and increasing rock strength.

As the intrinsic permeability of fault zones in mudstones is low, it is likely that faults do not generally enhance rates of fluid flow in mudstones. However, there is abundant evidence for episodic flow along faults in tectonically active regions such as accretionary complexes. There is also good evidence for fluid flow along faults in tectonically quiet sedimentary basins, but the timing and extent of enhanced permeability and enhanced flow is poorly constrained. Conditions conducive to fluid flow along muddy faults include the activity of the fault, high fluid pressures within the fault zone, and the extent of overconsolidation and lithification of the mudstones. Many more observations are required to quantify the importance of faults and fractures as fluid conduits in sedimentary basins. The current dataset (Neuzil 1994) suggests that fluid flow at the basinal (kilometre) scale may be perturbed only by widely spaced transmissive fractures rather than by episodic flow along small, active faults.

Our understanding of mudstone permeability is still relatively primitive. Some areas where improvements are required and feasible include:

(1) the establishment of a larger permeability database for well-characterized, natural mudstones, especially more deeply buried mudstones; these data should be used to calibrate existing permeability models;

(2) an analysis of the effect of scale on mudstone permeability, and determination of how to use data gathered on core plugs to estimate permeabilities of mudstone sequences;

(3) a more quantitative understanding of the role of faults and fractures as conduits for fluid flow both through and across mudstone sequences; this must include an understanding of the way in which fluids are delivered to the fault zone;

(4) a more complete description of the mechanical behaviour of mudstones and the way this changes with increased burial and chemical diagenesis; this will feed into predictive models of mudstone faulting and fracturing, and the role of these processes in fluid flow;

(5) a fuller understanding of petroleum migration through mudstones, including a description of relative permeability.

References

AHN, J. H. & PEACOR, D. R. 1986. Transmission and analytical electron microscopy of the smectite to illite transition. *Clays and Clay Minerals*, **34**, 165–179.

AL-TABBAA, A. & WOOD, D. M. 1987. Some measurements of the permeability of kaolin. *Géotechnique*, **37**, 499–503.

APLIN, A. C., YANG, Y. & HANSEN, S. 1995. Assessment of β, the compression coefficient of mudstones and its relationship with detailed lithology, *Marine and Petroleum Geology*, **12**, 955–963.

ARCH, J. & MALTMAN, A. J. 1990. Anisotropic permeability and tortuosity in deformed wet sediments. *Journal of Geophysical Research*, **95**, 9035–9047.

—— & —— 1993. Anisotropic permeability and tortuosity in deformed wet sediments: Reply. *Journal of Geophysical Research*, **98**, 17864–17865.

ATKINSON, J. H. & BRANSBY, P. L. 1978. *The Mechanics of Soils — An Introduction to Critical State Soil Mechanics*. McGraw-Hill, London.

BEHRMANN, J. H. 1991. Conditions for hydrofracture and fluid permeability of accretionary wedges. *Earth and Planetary Science Letters*, **107**, 550–558.

BJØRLYKKE, K. & HØEG, K. 1997. Effects of burial diagenesis on stresses, compaction and fluid flow in sedimentary basins. *Marine and Petroleum Geology*, **14**, 267–276.

BOLTON, A. & MALTMAN, A. J. 1997. Fluid flow pathways in actively deforming sediments: the role of pore fluid pressures and volume change. *In*: HENDRY, J. *et al.* (eds) *Geofluids II, Second International Conference on Fluid Evolution, Migration and Interaction in Sedimentary Basins and Orogenic Belts*, 356–359.

Borst, R. L. 1982. Some effects of compaction and geological time on the pore parameters of argillaceous rocks. *Sedimentology*, **29**, 291–298.

Brace, W. F. 1980. Permeability of crystalline and argillaceous rocks. *International Journal of Rock Mechanics and Mining Science*, **17**, 241–245.

——, Paulding, B. W. & Scholz, C. 1966. Dilatancy in the fracture of crystalline rocks. *Journal of Geophysical Research*, **71**, 3939–3953.

Bredehoeft, J. D., Neuzil, C. E. & Milly, P. C. D. 1983. *Regional Flow in the Dakota Aquifer: a Study of the Role of Confining Layers*. US Geological Survey Water Supply Paper, **2237**.

Bromhead, E. N. 1992. *The Stability of Slopes*, 2nd edn. Chapman and Hall, New York.

Brown, K. M. 1994. Fluids in deforming sediments. *In*: Maltman, A. J. (ed.) *Geological Deformation of Sediments*. Chapman and Hall, London, 205–237.

——, Bekins, B., Clennell, M. B., Dewhurst, D. N. & Westbrook, G. K. 1994. Heterogeneous hydrofracture development and accretionary fault dynamics. *Geology*, **22**, 259–262.

—— & Moore, J. C. 1993. A comment on Arch and Maltman 1990, 'Anisotropic permeability and tortuosity in deformed wet sediments'. *Journal of Geophysical Research*, **98**, 17859–17864.

Buckley, D. E. 1989a. Small fractures in deep sea sediments: indicators of pore fluid migration along compaction faults. *In*: Freeman, T. J. (ed.) *Disposal of Radioactive Waste in Seabed Sediments*. Graham and Trotman, London, 115–135.

—— 1989b. Geochemical evidence of pore water advection along a fault in plastic sediments from the southern Nares abyssal plain (western North Atlantic). *Chemical Geology*, **75**, 43–60.

—— & Grant, A. C. 1985. Fault-like features in abyssal plain sediments: possible dewatering structures. *Journal of Geophysical Research*, **90**, 9173–9180.

Capuano, R. M. 1993. Evidence of fluid flow in microfractures in geopressured shales. *Bulletin, American Association of Petroleum Geologists*, **77**, 1303–1314.

—— 1994. Evidence of fluid flow in microfractures in geopressured shales — reply. *Bulletin, American Association of Petroleum Geologists*, **78**, 1641–1646.

Carman, P. C. 1956. *The Flow of Gases through Porous Media*. Academic Press, New York.

Carson, B., Holmes, M. L., Umstatt, D., Strasser, J. C. & Johnson, H. P. 1991. Fluid expulsion from the Cascadia accretionary prism: evidence from porosity distribution, direct measurement and GLORIA imagery. *Philosophical Transactions of the Royal Society of London, Series A*, **335**, 331–340.

——, Westbrook, G. K., Musgrave, R. J. et al. 1995. *Proceedings of the Ocean Drilling Program, Scientific Results*, **146**. Ocean Drilling Program, College Station, TX.

Cartwright, J. A. 1994. Episodic basin-wide hydrofracturing of overpressured Early Cenozoic mudstone sequences in the North Sea Basin. *Marine and Petroleum Geology*, **11**, 587–607.

—— & Dewhurst, D. N. 1999. Layer-bound compaction faults in fine grained sediments. *Geological Society of America Bulletin*, **110**, 1242–1257.

Chu, C. L., Wang, C. Y. & Lin, W. 1981. Permeability and frictional properties of San Andreas fault gouges. *Geophysics Research Letters*, **8**, 565–568.

Clennell, M. B., Dewhurst, D. N., Brown, K. M. & Westbrook, G. K. 1999. Permeability anisotropy of consolidated clays. This volume.

——, Knipe, R. J. & Fisher, Q. J. 1998. Fault zones as barriers to or conduits for fluid flow in argillaceous formations: a microstructural and petrophysical perspective. *In: Fluid Flow through Faults and Fractures in Argillaceous Formations, Proceedings of OECD Conference on Clay Behaviour, Bern*, 125–139.

Darby, D., Haszeldine, R. S. & Couples, G. D. 1996. Pressure cells and pressure seals in the UK Central Graben. *Marine and Petroleum Geology*, **13**, 865–878.

Davis, S. N. 1988. Sandstones and Shales. *In*: Back, W., Rosenheim, J. S. & Seaber, P. R. (eds) *The Geology of North America, Vol. O-2, Hydrogeology*. Geological Society of America, Boulder, CO, 323–332.

Delage, P. & Lefebvre, G. 1984. Study of the structure of a sensitive Champlain clay and its evolution during consolidation. *Canadian Geotechnical Journal*, **21**, 21–35.

——, Tessier, D. & Marcel-Audiguier, M. 1982. Use of the Cryoscan apparatus for observation of freeze-fractured planes of a sensitive Quebec clay in scanning electron microscopy. *Canadian Geotechnical Journal*, **19**, 111–114.

Dewhurst, D. N., Aplin, A. C. & Sarda, J.-P. 1995. *Controls on the Permeability of Compacting Mudstones*. Report de l'Institut Français du Pétrole, **42555**.

——, ——, —— & Yang, Y. 1998a. Compaction-driven evolution of porosity and permeability in natural mudstones: an experimental study. *Journal of Geophysical Research*, **103**(B1), 651–661.

——, Brown, K. M., Clennell, M. B. & Westbrook, G. K. 1996a. A comparison of the fabric and permeability anisotropy of consolidated and sheared silty clay. *Engineering Geology*, **42**, 253–267.

——, Clennell, M. B., Brown, K. M. & Westbrook, G. K. 1996b. Fabric and hydraulic conductivity of sheared clays. *Geotechnique*, **46**, 761–769.

——, Westbrook, G. K., Clennell, M. B. & Brown, K. M. 1998b. Permeability anisotropy in clay-rich shear zones. *In: Fluid Flow through Faults and Fractures in Argillaceous Formations, Proceedings of OECD Conference on Clay Behaviour, Bern*, 287–298.

Dick, J. C., Shakoor, A. & Wells, N. 1994. A geological approach toward developing a mudstone durability classification system. *Canadian Geotechnical Journal*, **31**, 17–27.

DI MAIO, C. 1996. Exposure of bentonite to salt solution: osmotic and mechanical effects. *Géotechnique*, **46**, 695–707.

DROST-HANSEN, W. 1991. Some effects of vicinal water on the sedimentation process, compaction and ultimate properties of sediments. *In*: BENNETT, R. H., BRYANT, W. R. & HULBERT, M. H. (eds) *Microstructures of Fine Grained Sediments*. Springer, New York, 259–266.

FISHER, A. T. & HOUNSLOW, M. W. 1990. Heat flow through the toe of the Barbados accretionary complex. *In*: MOORE, J. C., MASCLE, A. *et al.* (eds) *Proceedings of the Ocean Drilling Program, Scientific Results*, **110**. Ocean Drilling Program, College Station, TX, 345–364.

FREEZE, R. A. & CHERRY, J. A. 1978. *Groundwater*. Prentice-Hall, Englewood Cliffs, NJ.

GAARENSTROOM, L., TROMP, R. A. J., DE JONG, M. C. & BRANDENBURG, A. M. 1993. Overpressures in the Central North Sea: implications for trap integrity and drilling safety. *In*: PARKER, J. R. (ed.) *Petroleum Geology of Northwest Europe: Proceedings of the 4th Conference*. Geological Society, London, 1305–1313.

GARCIA-BENGOCHEA, I. & LOVELL, C. W. 1981. Correlative measurements of pore size distribution and permeability in soils. *In*: ZIMMIE, T. F. & RIGGS, C. O. (eds) *Permeability and Groundwater Contaminant Transport*. American Society for Testing of Materials, Special Technical Publication, **746**, 137–150.

——, —— & ALTSCHAEFFL, A. G. 1979. Pore distribution and permeability of silty clays. *Journal of the Geotechnical Engineering Division*, ASCE, **105**, 839–856.

GIESKES, J. M., VROLIJK, P. & BLANC, G. 1990. Hydrogeochemistry, ODP Leg 110: an overview. *In*: MOORE, J. C., MASCLE, A. *et al.* (eds) *Proceedings of the Ocean Drilling Program, Scientific Results*, **110**. Ocean Drilling Program, College Station, TX, 395–408.

GRIFFITHS, F. J. & JOSHI, R. C. 1989. Change in pore size distribution due to consolidation of clays. *Geotechnique*, **39**, 159–167.

—— & —— 1990. Clay fabric response to consolidation. *Applied Clay Science*, **5**, 37–66.

GRIMM, K. A. & ORANGE, D. L. 1997. Synsedimentary fracturing, fluid migration, and subaqueous mass wasting: intrastratal microfractured zones in laminated diatomaceous sediments, Miocene Monterey Formation, California, USA. *Journal of Sedimentary Research*, **67**, 601–613.

HARDING, T. P. & TUMINAS, A. C. 1988. Interpretation of footwall (lowside) fault traps sealed by reverse faults and convergent wrench faults. *Bulletin, American Association of Petroleum Geologists*, **72**, 738–757.

—— & —— 1989. Structural interpretation of hydrocarbon traps sealed by basement normal blocks and at stable flank of foredeep basins and at rift basins. *Bulletin, American Association of Petroleum Geologists*, **73**, 812–840.

HELING, D. 1970. Microfabrics of shales and their rearrangement by compaction. *Sedimentology*, **17**, 247–260.

HILL, I. A., TAIRA, A., FIRTH, J. V. *et al.* 1993. *Proceedings of the Ocean Drilling Program, Scientific Results*, **131**. Ocean Drilling Program, College Station, TX.

HOOPER, E. C. D. 1991. Fluid migration along growth faults in compacting sediments. *Journal of Petroleum Geology*, **14**, 161–180.

HOWER, J., ESLINGER, E. V., HOWER, M. E. & PERRY, E. A. 1976. Mechanism of burial metamorphism of argillaceous sediment: 1. Mineralogical and chemical evidence. *Geological Society of America Bulletin*, **87**, 725–737.

HUBBERT, M. K. & RUBEY, W. W. 1959. Role of pore fluid pressures in the mechanics of overthrust faulting. *Geological Society of America Bulletin*, **70**, 115–205.

HUNT, J. M. 1990. Generation and migration of petroleum from abnormally pressured fluid compartments. *Bulletin, American Association of Petroleum Geologists*, **74**, 1–12.

INGRAM, G. M. & URAI, J. L. 1999. Top-seal leakage through faults and fractures: the role of mudstone properties. This volume.

IRWIN, H., CURTIS, C. & COLEMAN, M. 1977. Isotopic evidence for source of diagenetic carbonates formed during burial of organic-rich sediments. *Nature*, **269**, 209–213.

ISRAELACHVILI, J. N. 1992. *Intermolecular and Surface Forces*, 2nd edn. Academic Press, New York.

JONES, M. 1994. Mechanical principles of sediment deformation. *In*: MALTMAN, A. J. (ed.) *Geological Deformation of Sediments*. Chapman and Hall, London, 37–71.

JONES, M. E. & ADDIS, M. A. 1985. The application of stress-path and critical state analysis to sediment deformation. *Journal of Structural Geology*, **8**, 575–580.

JUANG, C. H. & HOLTZ, R. D. 1986a. A probabilistic permeability model and the pore size density function. *International Journal for Numerical and Analytical Methods in Geomechanics*, **10**, 543–553.

—— & —— 1986b. Fabric, pore size distribution and permeability of sandy soils. *Journal of Geotechnical Engineering*, **112**, 855–868.

KATSUBE, T. J. & WILLIAMSON, M. A. 1994. Effects of diagenesis on clay nanopore structure and implications for sealing capacity. *Clay Minerals*, **29**, 451–461.

—— & —— 1995. Critical depth of burial of subsiding shales and its effect on abnormal pressure development. *In*: BELL, J. S., BIRD, T. D., HILLIER, T. L. & GREENER, P. L. (eds) *Proceedings of the Oil and Gas Forum '95, Energy from Sediments*. Geological Survey of Canada Open File, **3058**.

KELLER, C. K., VAN DER KAMP, G. & CHERRY, J. A. 1989. A multiscale study of the permeability of a thick clayey till. *Water Resources Research*, **25**, 2299–2317.

KNIPE, R. J. 1986. Deformation path mechanisms for sediments undergoing lithification. *In*: MOORE, J. C. (ed.) *Structural Fabric in Deep Sea Drilling Project Cores from Forearcs*. Geological Society of America, Memoir, **166**, 151–160.

—— 1993. Micromechanisms of deformation and fluid flow behaviour during faulting. *In*: HICKMAN, S., SIBSON, R. H. & BRUHN, R. (eds) *The Mechanical Behaviour of Fluids in Fault Zones*. US Geological Survey Open File Report **94-228**, 301–310.

—— 1997. Juxtaposition and seal diagrams to help analyse fault seals in hydrocarbon reservoirs. *Bulletin, American Association of Petroleum Geologists*, **81**, 187–195.

LAMBE, T. W. & WHITMAN, R. V. 1979. *Soil Mechanics, SI version*. Wiley, New York.

LAPIERRE, C., LEROUEIL, S. & LOCAT, J. 1990. Mercury intrusion and permeability of Louiseville clay. *Canadian Geotechnical Journal*, **27**, 761–773.

LE PICHON, X., KOBAYASHI, K. *et al.* 1992. Fluid venting activity within the eastern Nankai Trough accretionary wedge: a summary of the 1989 Kaiko–Nankai results. *Earth and Planetary Science Letters*, **109**, 303–318.

LEONARDS, G. H. 1962. *Engineering Properties of Soils*. McGraw-Hill, New York.

LEROUEIL, S., BOUCLIN, G., TAVENAS, F., BERGERON, I. & LA ROCHELLE, P. 1990. Permeability anisotropy of natural clays as a function of strain. *Canadian Geotechnical Journal*, **27**, 568–579.

LEWIS, S. D., BEHRMANN, J. H., MUSGRAVE, R. J. *et al.* 1995. *Proceedings of the Ocean Drilling Program, Scientific Results*, **141**. Ocean Drilling Program, College Station, TX.

LOGAN, J. M. 1992. The influence of fluid flow on the mechanical behaviour of faults. *In*: TILLERSON, J. R. & WAWERSWIK, W. R. (eds) *Proceedings of the 33rd US Symposium on Rock Mechanics*, 141–150.

LORENZ, J. C., TEUFEL, L. W. & WARPINSKI, N. R. 1991. Regional fractures I: A mechanism for the formation of regional fractures at depth in flat-lying reservoirs. *Bulletin, American Association of Petroleum Geologists*, **75**, 1714–1737.

LUO, X. & VASSEUR, G. 1996. Geopressuring mechanism of organic matter cracking; numerical modelling. *Bulletin, American Association of Petroleum Geologists*, **80**, 856–873.

LUPINI, J. F., SKINNER, A. E. & VAUGHAN, P. R. 1981. The drained residual strength of cohesive soils. *Geotechnique*, **31**, 181–213.

MALTMAN, A. J. 1988. The importance of shear zones in naturally deformed wet sediments. *Tectonophysics*, **145**, 163–175.

MANN, D. M. & MACKENZIE, A. S. 1990. Prediction of pore fluid pressure in sedimentary basins. *Marine and Petroleum Geology*, **7**, 55–65.

MARSHALL, T. J. 1958. A relation between permeability and size distribution of pores. *Journal of Soil Science*, **9**, 1–8.

MAUBEUGE, F. & LERCHE, I. 1994. Geopressure evolution and hydrocarbon generation in a north Indonesian basin: two-dimensional quantative modelling. *Marine and Petroleum Geology*, **11**, 104–115.

MESRI, G. & OLSON, R. E. 1971. Mechanisms controlling the permeability of clays. *Clays and Clay Minerals*, **19**, 151–158.

MILLIKEN, K. L. & LAND, L. S. 1994. Evidence of fluid flow in microfractures in geopressured shales — discussion. *Bulletin, American Association of Petroleum Geologists*, **78**, 1637–1640.

MITCHELL, J. K. 1993. *Fundamentals of Soil Behaviour*, 2nd edn. Wiley, New York.

——, HOOPER, D. R. & CAMPANELLA, R. G. 1965. Permeability of compacted clay. *Journal of the Soil Mechanics and Foundation Division*, ASCE, **91**, 41–65.

MOORE, J. C. 1989. Tectonics and hydrogeology of accretionary prisms: role of the décollement zone. *Journal of Structural Geology*, **11**, 95–106.

——, MASCLE, A. *et al.* 1988. Tectonics and hydrogeology of the northern Barbados Ridge: results from Ocean Drilling Program Leg 110. *Geological Society of America Bulletin*, **100**, 1578–1593.

——, —— *et al.* 1990. *Proceedings of the Ocean Drilling Program, Scientific Results*, **110**. Ocean Drilling Program, College Station, TX.

——, SHIPLEY, T. H., GOLDBERG, D. *et al.* 1995. Abnormal fluid pressures and fault zone dilation in the Barbados accretionary prism: evidence from logging while drilling. *Geology*, **23**, 605–608.

MORGENSTERN, N. R. & TCHALENKO, J. S. 1967. Microscopic structures in kaolin subjected to direct shear. *Geotechnique*, **17**, 309–328.

MORROW, C., SHI, L. Q. & BYERLEE, J. 1981. Permeability and strength of fault gouge under high pressure. *Geophysics Research Letters*, **8**, 325–328.

——, —— & —— 1984. Permeability of fault gouge under confining pressure and shear stress. *Journal of Geophysical Research*, **89**, 3193–3200.

NAGARAJ, T. S., VATSALA, A. & SRINIVASA MURTHI, B. R. 1990. Discussion of Griffiths & Joshi 'Change in pore size distribution due to consolidation of clays'. *Géotechnique*, **40**, 303–309.

NEUZIL, C. E. 1986. Groundwater flow in low-permeability environments. *Water Resources Research*, **22**, 1163–1195.

—— 1993. Low fluid pressure in the Pierre Shale: a transient response to erosion. *Water Resources Research*, **29**, 2007–2020.

—— 1994. How permeable are clays and shales? *Water Resources Research*, **30**, 145–150.

NEWMAN, A. C. D. 1987. *Chemistry of Clays and Clay Minerals*. Mineralogical Society Monograph, **6**.

OLSEN, H. W. 1962. Hydraulic flow through saturated clays. *Clays and Clay Minerals*, **9**, 131–162.

—— 1965. Deviations from Darcy's Law in saturated clays. *Proceedings of the American Soil Science Society*, **29**, 135–140.

OSBORNE, M. J. & SWARBRICK, R. E. 1997. Mechanisms for generating overpressure in sedimentary basins: a re-evaluation. *Bulletin, American Association of Petroleum Geologists*, **81**, 1023–1041.

ROBERTS, S. J. & NUNN, J. A. 1995. Episodic fluid expulsion from geopressured sediments. *Marine and Petroleum Geology*, **12**, 195–204.

RUDOLPH, D. L., CHERRY, J. A. & FARVOLDEN, R. N. 1991. Groundwater flow and solute transport in fractured lacustrine clay near Mexico City. *Water Resources Research*, **27**, 2187–2201.

RYE-LARSEN, M. 1994. The Balder Field: refined reservoir interpretation with the aid of high resolution seismic data and seismic attribute mapping. *In*: AASEN, J. O. *et al.* (eds) *North Sea Oil and Gas Reservoirs, III*. Kluwer, Dordrecht, 115–124.

SCHEIDEGGER, A. E. 1974. *The Physics of Flow through Porous Media*, 3rd edn. University of Toronto Press.

SCHLÖMER, S. & KROOSS, B. M. 1997. Experimental characterisation of the hydrocarbon sealing efficiency of caprocks. *Marine and Petroleum Geology*, **14**, 565–580.

SCHNEIDER, F., BURRUS, J. & WOLF, S. 1993. Modelling overpressures by effective-stress/porosity relationships in low-permeability rocks: empirical artifice of physical reality? *In*: DORÉ, A. G. (ed.) *Basin Modelling: Advances and Applications*. Norwegian Petroleum Society (NPF) Special Publication, **3**, 333–341.

SCHOFIELD, A. N. & WROTH, C. P. 1968. *Critical State Soil Mechanics*. McGraw-Hill, London.

SCREATON, E. J., WUTHRICH, D. R. & DREISS, S. J. 1990. Fluid flow within the Barbados Ridge complex, part 1: Dewatering in the toe of the prism. *In*: MOORE, J. C., MASCLE, A. *et al.* (eds) *Proceedings of the Ocean Drilling Program, Scientific Results*, **110**. Ocean Drilling Program, College Station, TX, 321–329.

SHIPLEY, T. H., MOORE, G. F., BANGS, N. L., MOORE, J. C. & STOFFA, P. L. 1994. Seismically inferred dilatancy distribution, northern Barbados ridge décollement: implications for fluid migration and fault strength. *Geology*, **22**, 411–414.

SIBSON, R. H. 1981. Controls on low stress hydrofracture dilatancy in thrust, wrench and normal fault terrains. *Nature*, **289**, 665–667.

——, MOORE, J. M. & RANKIN, A. H. 1975. Seismic pumping: a hydrothermal fluid transport mechanism. *Journal of the Geological Society, London*, **131**, 653–659.

SKEMPTON, A. W. 1964. Long term stability of clay slopes. *Géotechnique*, **14**, 75–101.

—— 1985. Residual strength of clays in landslides, folded strata and the laboratory. *Géotechnique*, **35**, 3–18.

SMITH, D. A. 1980. Sealing and nonsealing faults in Louisiana Gulf coast salt basin. *Bulletin, American Association of Petroleum Geologists*, **64**, 145–172.

STEPHENSON, E. L., MALTMAN, A. J. & KNIPE, R. J. 1994. Fluid flow in actively deforming sediments; dynamic permeability in accretionary prisms. *In*: PARNELL, J. (ed.) *Geofluids: Origin, Migration and Evolution of Fluids in Sedimentary Basins*. Geological Society, London, Special Publications, **78**, 113–125.

TAVENAS, F., JEAN, P., LEBLOND, P. & LEROUEIL, S. 1983*a*. The permeability of natural soft clays. Part 2: Permeability characteristics. *Canadian Geotechnical Journal*, **20**, 645–660.

——, LEBLOND, P., JEAN, P. & LEROUEIL, S. 1983*b*. The permeability of natural soft clays. Part 1: Methods of laboratory measurement. *Canadian Geotechnical Journal*, **20**, 629–644.

TOKUNAGA, T., HOSOYA, S., KOJIMA, K. & TOSAKA, H. 1994. Change of hydraulic properties of muddy deposits during compaction: assessment of mechanical and chemical effect. *Proceedings of the International Association of Engineering Geologists Conference*, **7**, 635–643.

VAN OLPHEN, H. 1977. *An Introduction to Clay Colloid Chemistry*. John Wiley & Sons, New York.

VERSCHUREN, M. 1992. *An integrated 3D approach to clay tectonic deformation*. PhD thesis, University of Ghent, Belgium.

VROLIJK, P. 1987. Tectonically driven fluid flow in the Kodiak Accretionary Complex, Alaska. *Geology*, **15**, 466–469.

WARD, C. D., COGHILL, K. & BROUSSARD, M. D. 1994. *The Application of Petrophysical Data to Improve Pore and Fracture Pressure Determination in North Sea Central Graben HPHT Wells*. Society of Petroleum Engineers Paper, **28297**.

YANG, Y. & APLIN, A. C. 1998. Influence of lithology and compaction on the pore size distribution and modelled permeability of some mudstones from the Norwegian margin. *Marine and Petroleum Geology*, **15**, 163–175.

ZNIDARCIC, D. & AIBAN, S. A. 1988. Comment on Al-Tabbaa and Wood: 'Some measurements of the permeability of kaolin'. *Géotechnique*, **38**, 453–454.

Thermal conductivity of sedimentary rocks: uncertainties in measurement and modelling

KIRSTI MIDTTØMME & ELEN ROALDSET

Department of Geology and Mineral Resources Engineering, Norwegian University of Science and Technology (NTNU), 7491 Trondheim, Norway

Abstract: In spite of the fact that thermal conductivity is a key factor in basin modelling, knowledge regarding the thermal conductivity of sedimentary rocks is scarce. In particular, hardly any information exists about claystones and mudstones which can make up 70–80% of a sedimentary basin. For the Upper Jurassic sediments of a section across the northern North Sea, basin modelling programs, using the geometric mean model, gave a deviation of 50°C for the present-day temperature, simply because the matrix conductivity was estimated by two different models. These uncertainties in the determination of thermal conductivities are partly due to problems in measuring, and partly to difficulties in modelling the thermal conductivity. Future work should concentrate on comparative studies and tests on all types of sedimentary rocks, directed towards improved measurement methods. This is of particular importance for clays, mudstones and shales, as measurements on these are associated with the greatest uncertainties. Standardized procedures that include detailed descriptions of sampling, sample preparation and measurement techniques are important to ensure that the thermal conductivity measured is unaffected by factors related to the measurement method. In thermal conductivity modelling it is our experience that the influence of sediment texture is underestimated by most models currently applied. Estimates based on the geometric mean model seem to be most reliable when the determination of matrix conductivity is restricted to the sediment type, mineralogy and texture of the sediments.

Thermal conductivity, k, is a key variable in thermal modelling, because it controls the temperature within sedimentary basins. The temperature gradient as a result of conduction is described by Fourier's law as inversely proportional to the thermal conductivity for a given heat flow:

$$q = -k\frac{dT}{dz} \quad (1)$$

where q is heat flow (W m^{-2}), k is thermal conductivity (W m^{-1} K^{-1}) and dT/dz is temperature gradient.

In 1989 Blackwell & Steele concluded that information is too sparse to estimate mean thermal conductivity effectively for a section of sedimentary rocks and, if the mean conductivity cannot be accurately predicted, even the most sophisticated and appropriate modelling techniques are not sufficient for accurate temperature predictions. There still exists a lack of basic knowledge about the thermal conductivity of sedimentary rocks and, in particular, reliable information concerning mudstones and shales is scarce (e.g. Gallagher *et al.* 1997). In basin modelling it is still unclear which variables affect the thermal conductivity. Until the correct variables are found, attempts to develop thermal conductivity models without clear limitations are difficult. Laboratory measurements of thermal conductivity form the basis for thermal conductivity modelling of sedimentary basins. Lack of standardized procedures and quality control seems to have been the cause of the variable quality of some published thermal conductivity measurements (see discussion by Midttømme *et al.* 1997*a*). According to Brigaud & Vasseur (1989), published data on thermal conductivities of clays are generally unreliable because no distinctions are made on the mineralogy and saturation conditions of the experiments.

This paper summarizes the main results obtained from the project 'Shale and Claystones, Physical and Mineralogical Parameters in Basin Modelling' financed by the Norwegian Research Council and Statoil. On the basis of our experience of thermal conductivity studies and measurements, mainly on mudstones, we will discuss problems and suggest how improvements

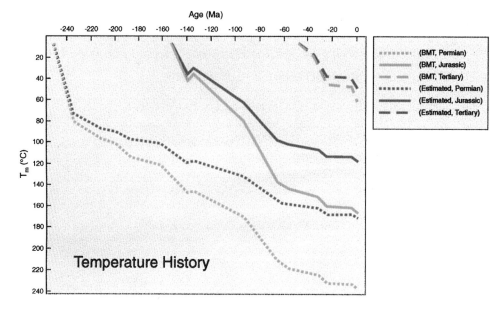

Fig. 1. Temperature histories of three sediments (Permian, Jurassic and Tertiary) in the North Sea Basin. Thermal conductivities are estimated by two methods: (1) estimates by the geometric mean model from the mineral conductivities (Table 1), shown with red lines; (2) estimates by the geometric mean based on the BMT database of thermal conductivity measurements and mineralogy, shown with green lines.

may be made in the measurement and modelling of thermal conductivities.

Example from the North Sea

The importance of thermal conductivity on modelled temperature history is illustrated in Fig. 1. The temperature history through a section across the northern North Sea is estimated using BMT (Basin Modelling Toolbox), a basin modelling system developed by Rogaland Research and described by Fjeldskaar et al. (1990). A database of measured thermal conductivities is included in the program.

Thermal conductivity is estimated by a model given by the statistical expression geometric mean:

$$k = k_f^\phi k_s^{(1-\phi)} \qquad (2)$$

where k is thermal conductivity of the bulk sediment (W m^{-1} K^{-1}), k_f is thermal conductivity of the pore fluid (W m^{-1} K^{-1}), k_s is thermal conductivity of the matrix (the solid part of the sediments, i.e. mineral grains and solid organic matters) (W m^{-1} K^{-1}) and ϕ is porosity (0.00–1.00).

The matrix conductivity (k_s) in this model is determined by two methods: (1) by the geometric mean model from the mineralogy based on mineral conductivities (Table 1) ('Estimated' in Fig. 1); (2) from a BMT database of thermal conductivity measurements based on the mineralogy ('BMT' in Fig. 1). The estimated temperature histories of three selected points in

Table 1. Thermal conductivity of common minerals and fluids in sedimentary rocks

Mineral	Thermal conductivity (W m^{-1} K^{-1})	Reference
Quartz	7.8	Horai 1971
Calcite	3.4	Horai 1971
Dolomite	5.1	Horai 1971
Anhydrite	6.4	Horai 1971
Pyrite	19.2	Horai 1971
Siderite	3.0	Horai 1971
K-feldspar	2.3	Horai 1971
Albite	2.3	Horai 1971
Mica	2.3	Horai 1971
Halite	6.5	Horai 1971
Kaolinite	2.8	Horai 1971
Illite	1.8	Horai 1971
Mixed layer I/S	1.9	Horai 1971
Water (20°C)	0.60	Schön 1996
Oil	0.21	Jensen & Doré 1993
Gas	0.079	Jensen & Doré 1993
Air (20°C)	0.026	Schön 1996

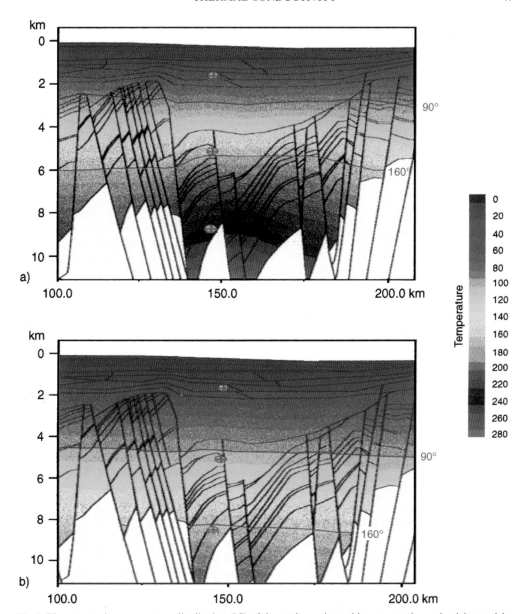

Fig. 2. The present-day temperature distribution (°C) of the section estimated by two matrix conductivity models: (a) matrix conductivity is determined from the BMT database using mineralogical information; (b) matrix conductivity is estimated from the mineralogy based on mineral conductivities. The three points represent Lower Tertiary, Upper Jurassic and Permian values. The red lines mark the oil window (90–160°C).

the cross-section representing different geological ages are shown in Fig. 1. The points represent Lower Tertiary (depth 1650 m b.s.b.), Upper Jurassic (depth 5150 m b.s.b.) and Permian sediments (depth 8600 m b.s.b.). The thermal conductivities are estimated in both models by the geometric mean model by using the same porosity values. The present-day temperatures estimated by using the two methods are shown in Fig. 2. By applying the thermal conductivity database to estimate matrix conductivity, the present-day temperature for the Permian sediments is estimated to be about 70°C higher than the values obtained when using the geometric mean model to estimate matrix conductivity. For the Lower Tertiary sediments, a discrepancy of over 10°C is estimated for the present-day temperature modelled by the two models.

Measurement of thermal conductivity

Measured thermal conductivities of sedimentary rocks have been presented in several handbooks and papers (e.g. Clark 1966; Kappelmeyer & Hanel 1974; Schön 1996) (Table 2). Whereas Clark (1966) tabulated the thermal conductivities of sandstones in the range of 1.46–4.27 W $m^{-1} K^{-1}$ based on 26 measurements, Schön (1996) included 1720 measurements of sandstones ranging from 0.38 to 6.50 W $m^{-1} K^{-1}$. Except for a larger range, the increased number of thermal conductivity values has so far not really improved our knowledge of the thermal conductivity of sedimentary rocks. Poor quality measurements are assumed to be the main reason for the uncertainty related to the thermal conductivity values, and are assumed to have caused many unreliable values, especially for shales and mudstones.

Methods of measurement

There is no standard method for measuring the thermal conductivity of sedimentary rocks. Different measurement methods have been developed, the two main techniques being the *divided bar method* and the *needle probe method*. The former is a steady-state method where the thermal conductivity is determined by Fourier's law (equation (1)) when there is a constant temperature gradient across the samples and the heat flow is stable. Different designs of the divided bar apparatus have been developed. This method is considered to be the most accurate one (Johansen 1975; Farouki 1981; Brigaud *et al.* 1990) and is recommended by the International Heat Flow Commission to be applied wherever possible for competent rock

Table 2. *Thermal conductivities of sedimentary rocks*

(Clark 1966)

Rock type	Locality	Thermal conductivity (W $m^{-1} K^{-1}$)			Reference
		n	Range	Mean	
Sandstone	Karoo Sandstone	7	1.46–3.22	1.97	Mossop & Gafner (in Clark 1966)
	Jacobsville Sandstone	8	2.13–4.27	2.83	Birch 1954
	Carboniferous Sandstone	6	2.51–3.22	2.77	Bullard & Niblett (in Clark 1966)
Shale	Karoo Shale	6	1.97–2.87	2.38	Mossop & Gafner (in Clark 1966)
	Berry No. 1 Well, Kern Co., CA 1000–5290 ft	14	1.17–1.76	1.49	Benfield 1947
	Berry No. 1 Well, Kern Co., CA 5290–8780 ft	17	1.34–2.34	1.76	Benfield 1947
	Carboniferous Shale	11	1.26–1.80	1.36	Bullard & Niblett (in Clark 1966)

(Schön 1996)

Rock type	Thermal conductivity (W $m^{-1} K^{-1}$)			Reference
	n	Range	Mean	
Sandstone	1262	0.90–6.50	2.47	Cermak & Rybach 1982
	11	1.88–4.98	3.72	Jessop 1990
	447	0.38–5.17	1.66	Dortman (in Schön 1996) Kobranova (in Schön 1996)
Siltstone	3	2.47–2.84	2.68	Jessop 1990
Clay–siltstone	19	1.70–3.40	2.46	Cermak & Rybach 1982
Claystone	242	0.60–4.00	2.04	Cermak & Rybach 1982
Shale	377	0.55–4.25	2.07	Cermak & Rybach 1982

samples (Beck 1988). The needle probe method is a transient method originally developed by Von Herzen & Maxwell (1959). By this method a probe, which consists of a heating wire and a thermistor, is inserted into the sample. The temperature of the sample is recorded as a function of time. From the temperature–time curve, the thermal conductivity of the material can be determined (Jessop 1990). This method is simpler and far more rapid than the divided bar method, and fewer demands are made on sample preparation. The needle probe equipment is more standardized than that used in the divided bar method.

Good contact between the samples and the laboratory equipment is important to prevent contact resistance, an important source of error. Attaining satisfactory contact depends on the type and shape of the sample and the design of the equipment. A smooth and well-prepared sample surface that exactly fits the sample chamber is particularly important for the divided bar method. Different types of waxes and fluids are used to improve the sample contact during measurements. The different designs of the equipment make them suitable for different types of samples. A divided bar apparatus tested and calibrated for consolidated samples might therefore be carefully used for soft, unconsolidated samples.

Sampling, preparation and measurements

Even more important than the measurement method is the sample preparation. The sampling and sample preparation are critical for the measurement of mudstone and shale samples in particular. With fine-grained sediments the material has a tendency to break up in the sampling process and, even if the sampling is successful, the material may be subjected to volumetric expansion and/or desiccation under laboratory conditions (Brigaud & Vasseur 1989). This problem has resulted in fewer (Table 2) and probably poorer thermal conductivity measurements of mudstones and shales. Thermal conductivity of sedimentary rocks is measured, in general, on water-saturated samples. The very low thermal conductivity of air (Table 1) makes the thermal conductivity of sedimentary rocks very sensitive to drying of the samples during the process of sampling, preparation and measurement.

All methods of thermal conductivity measurement require heat transfer through the sample during measurement. This may lead to an unstable condition, as heat flow induces mass flow which in turn makes a convective contribution to the total heat flow and, to some degree, will also cause drying of the sample. The measurement error caused by fluid flow is reduced by maintaining a low temperature gradient across the samples during measurement. The risk of drying will also depend on the mean temperature of the samples during measurement. Measurements at high temperature will be more sensitive to drying.

The preparation method developed by Sass *et al.* (1971), where the samples are crushed to powder, and the measurements are carried out on a suspension of powdered samples and water, represents the extreme case. This approach has, however, been used for the divided bar method (Sass *et al.* 1971), the needle probe method (Horai 1971) and other transient methods (Middleton 1994). One advantage in measuring powdered samples is that it simplifies the sampling and the preparation process. Measurements can also be carried out on rock fragments such as drill-bit cuttings. The main drawback of this method is that it does not account for the possible effect of sediment texture.

The anisotropic and inhomogeneous nature of the sample

The anisotropic nature of sedimentary rocks is a common feature that affects their thermal conductivity. Thermal conductivities measured parallel ($k_{\parallel}$) to the layering appear in some cases to be more than twice those measured perpendicular ($k_{\perp}$) to it (e.g. Midttømme *et al.* 1996; Schön 1996). The significance of the direction of heat flow on the measurements highlights the importance of sample handling and measurement techniques, particularly for anisotropic rocks such as shale. The heterogeneous nature of most sedimentary rocks also introduces uncertainty in the estimation of the thermal conductivity of rock volumes used in basin models. A question that arises is how representative a sample of $1–10\,cm^3$ is for the whole sequence of rocks from which it was selected (Gallagher *et al.* 1997).

Measurements on parallel samples are presented in Table 3. All parallel samples, except those from the Kimmeridge Clay, were prepared from the same block with a distance between the parallel samples of <20 cm. Kimmeridge Clay samples come from two sites, A and B, located about 100 m apart in the same clay pit. The deviation for the two sites of Kimmeridge Clay samples is ±17.8%, and the internal deviations for sites A and B are ±1.5% and ±0.5%,

Table 3. *Thermal conductivity measurements on parallel prepared samples*

Sample	Measured thermal conductivity (W m^{-1} K^{-1})	Mean (W m^{-1} K^{-1})	Deviation (%)
Claystones and mudstones from England (Midttømme et al. 1998a)			
Fuller's Earth, Reigate	0.74/0.74/0.76	0.75	±1.8
Fuller's Earth, Baulking	0.66/0.68/0.71	0.68	±3.9
Kimmeridge Clay	0.68^A/0.70^A/0.96^B/0.97^B	0.83 (0.69^A/0.97^B)	±17.8 (±1.5^A/±0.5^B)
Oxford Clay	0.75/0.76/0.84	0.78	±4.3
London Clay	0.83/0.84	0.84	±0.6
Argillaceous sediments from Norwegian Continental Shelf (Midttømme et al. 1997)			
Upper Jurassic Heather Formation	0.93/0.97	0.95	±2.1
Upper Cretaceous	0.92/1.04	0.98	±6.1

Superscript A and B indicate two sites for Kimmeridge Clay.

respectively. When the variation in measured thermal conductivity is nearly 20% for the Kimmeridge Clay sampled at the same depth and 100 m apart, how representative are these measurements for Kimmeridge Clay samples taken 100–1000 km away and at other burial depths? Our present knowledge of thermal conductivity is too limited to allow such questions to be answered.

Comparison between needle probe and divided bar methods

Comparisons of measurements by the needle probe and the divided bar methods have previously been reported. Agreement was obtained by Von Herzen & Maxwell (1959), Sass et al. (1971) and Brigaud & Vasseur (1989), whereas higher values of thermal conductivity were measured with the needle probe by Penner (1963), Slusarchuk & Foulger (1973), Johansen (in Farouki, 1981), Somerton (1992) and Midttømme et al. (1998b). The discrepancy in these studies is of the order of 10–20%. A constant deviation between the two methods was measured for unconsolidated sediments (Midttømme et al. 1998b) (Fig. 3). This deviation was inferred to be due to a calibration error between the two methods of measurement, as other measurement errors caused by, for example, drying or disturbance of the samples,

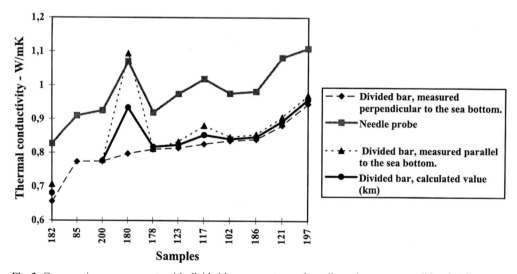

Fig. 3. Comparative measurements with divided bar apparatus and needle probe on unconsolidated sediments (Midttømme et al. 1998b). The divided bar measurements are carried out parallel and perpendicular to the sea bottom. To compare those measurements with the needle probe measurements a mean value, k_m, is calculated based on the equation for an ellipse ($k_m = \sqrt{(k_\perp \times k_\parallel)}$).

Table 4. *Comparative measurements on clay and mudstones from England (Midttømme et al. 1998a)*

Sample	Thermal conductivity (W m^{-1} K^{-1})			
	Divided bar apparatus, NTNU	Needle probe,* University of Aarhus	Middleton's method	Bloomer's measurement (1981)
London Clay,				
perpendicular	0.83	0.84 (1)		2.45 ± 0.07
parallel	1.19	0.94 (3)		NP† (5)‡
Fuller's Earth,				
perpendicular	0.68	0.73 (3)		1.95 ± 0.05
parallel	0.80	0.98 (1)		NP/PB (41)
Kimmeridge Clay,				
perpendicular	0.97	1.21 (1)	0.89/0.96	1.51 ± 0.09
parallel	1.20	1.21 (1)	1.18/1.07	NP/PB (58)
Oxford Clay,				
perpendicular	0.79	1.19 (2)	0.84	1.57 ± 0.03
parallel	1.11	1.29 (3)	1.11	PB (11)

*The number in parenthesis gives the quality of the measurements; one is good and three is poor.
†Method of measurements: NP, needle probe; PB, divided bar.
‡Number of measurements.

were assumed to cause a more random deviation. Comparative measurements carried out by the divided bar apparatus, the needle probe and the transient method of Middleton (1994) (Table 4) were reported in a study of clays and mudstones from England (Midttømme et al. 1998a). Previous measurements on these clays have also been published by Bloomer (1981), who applied both the needle probe and divided bar technique. Two main discrepancies can be recognized between the measurements of Bloomer (1981) and those of Midttømme et al. (1998a) (see Table 4): Bloomer's measurements are considerably higher than those of Midttømme et al. (1998a). Of the measured discrepancies, the deviation of over 100% for Fuller's Earth is most disquieting, as this clay appears to be nearly homogeneous and isotropic. This discrepancy can therefore hardly be due to mineralogical or textural variations of the investigated samples.

Of the three measurements published by Midttømme et al. (1998a), the needle probe measurements differ from the other two. No systematic discrepancy is found between the needle probe and the divided bar apparatus measurements, but for the three layered mudstones (London Clay, Kimmeridge Clay and Oxford Clay) a considerably lower thermal conductivity anisotropy ($a = k_{\parallel}/k_{\perp}$) is obtained by the needle probe than by the other two methods. This discrepancy in the measured anisotropy can be explained by the way heat is transferred through the samples by the different methods. In the needle probe method heat is transferred from a line source, whereas by the divided bar apparatus and Middleton's method it is transferred from an area.

In light of today's knowledge, more important than new thermal conductivity measurements are tests and comparative studies of the methods and procedures of measurements on all types of sedimentary rocks, and in particular on clays, mudstones and shales, as these measurements are associated with the greatest uncertainties. More standardized procedures and documentation of sampling, preparation and measurement are important to ensure that the measured thermal conductivity is not affected by factors related to the measurement method.

Modelling of thermal conductivity

Thermal conductivity in general has to be estimated without constraints from specific measurements in basin modelling. The accuracy of the modelled results depends on the lithological information available, the accuracy of this information and the knowledge of how the rock characteristics affect the thermal conductivity. The thermal conductivity of water-saturated sediments depends mainly on porosity, mineralogy and texture, and to some extent on temperature and pressure (e.g. Blackwell & Steele 1989; Midttømme et al. 1997). We now discuss the factors affecting thermal conductivity and give a review of different thermal conductivity models.

Porosity

A clear correlation is observed between decreasing porosity and increasing thermal conductivity in sedimentary rocks (e.g. Brigaud & Vasseur 1989). In a study of unconsolidated fine-grained sediments, thermal conductivity can be estimated within 80% accuracy from the water content (Midttømme et al. 1998b). Studies with no clear correlation between the two variables have been published for consolidated sediments (e.g. Blackwell & Steele 1989; Midttømme et al. 1997). Although the porosity is an important and clearly defined variable, it is not easy to determine accurately. Porosity can be determined from laboratory measurements, well-log data or compaction models. There seems to be some confusion among researchers with regard to the different types of porosities; e.g. effective porosity, total porosity and water content (Griffiths et al. 1992; Midttømme et al. 1998a). This confusion will be a source of error in the determination of thermal conductivity for some sediment types. In one example, the thermal conductivity of London Clay was estimated to be 21% higher when using porosity instead of water content in the thermal conductivity model (Midttømme et al. 1998a).

The large variation of the fluid conductivities (Table 1) makes the pore fluid an important factor controlling the thermal conductivity (e.g. Somerton 1992; Jensen & Doré 1993). In this study we have only considered water-saturated samples. The influence of other pore fluids is therefore ignored.

Mineralogy

As minerals have different thermal conductivities (Table 1), the composition of the matrix (solid part of the rock) will affect the thermal conductivity. The quartz content is considered to be most important, as quartz has a relatively high conductivity (e.g. Johansen 1975; Somerton 1992; McKenna et al. 1996). Thermal conductivities measured on fine-grained sediments such as clay, claystone and shale are, in general, lower than for coarser material, and these low values have been related to a higher content of clay minerals in these samples (Gilliam & Morgan 1987; Demongodin et al. 1991; Midttømme et al. 1997). The clay mineral content is also important, as the ability of these minerals to transfer heat seems to be more varied than for the other minerals (Midttømme & Roaldset 1997).

Texture

The effect of sediment texture on thermal conductivity is more complex and more complicated to model than the effect of mineralogy and porosity. The effect of texture is, in many studies, considered as less important than the porosity and mineralogy (e.g. Brigaud & Vasseur 1989; Somerton 1992; McKenna et al. 1996; Gallagher et al. 1997).

Anisotropy. Evidence of a textural effect is the measured anisotropy. Thermal conductivities parallel to the layering ($k_{\parallel}$) are, in certain cases, found to be more than double those perpendicular to it ($k_{\perp}$) (Midttømme et al. 1996; Schön 1996). Schön (1996) assumed three causes for this anisotropy: (1) crystal anisotropy of the individual rock-forming minerals; (2) intrinsic or structural anisotropy resulting from the mineral shapes and their textural arrangement within the rock; (3) orientation and geometry of cracks, fractures and other defects. The anisotropy of thermal conductivity ($a = k_{\parallel}/k_{\perp}$) will, from this assumption, be a function of burial history, depositional environment and mineralogy, mainly the content of clay minerals, as these have the greatest anisotropy (Table 5).

Grain size. A correlation is observed between grain size and thermal conductivity. Midttømme et al. (1998a) measured an increase in the thermal conductivity with increasing sand content of clay samples (Fig. 4). In a study of artificial samples of different grain size fractions of quartz, a linear correlation between the thermal conductivity and the logarithm of the grain size was observed (Midttømme & Roaldset 1998) (Fig. 5). A similar logarithmic correlation was observed for certain rocks by Rzhevsky & Norvik (1971). They suggested that the decrease in thermal conductivity with decreasing grain size was due to the increasing number of grain contacts per unit path of heat flow. For unconsolidated samples the grain size fractions, in particular the sand fraction, were found to have a strong effect on thermal conductivity, even greater than the complete mineralogy (Midttømme et al. 1998b).

Other textural factors. A continuing problem is how to include information on textural characteristics in a thermal conductivity model, as most of these are difficult to quantify. Nevertheless, textural factors, such as grain shape, grain arrangement, pore shape and pore size will to

Table 5. *Published anisotropy of thermal conductivity of some minerals*

Mineral	Symmetry	$k_\perp$ (W m^{-1} K^{-1})	$k_\parallel$ (W m^{-1} K^{-1})	Anisotropy ($a = k_\parallel/k_\perp$)	Reference
Muscovite	monoclinic	0.84	5.1	6.1	Cermak & Rybach 1982
Phlogopite (mica)		1.7	4.9	2.9	Goldsmid & Bowley 1960
Orthoclase	monoclinic	2.9	4.6	1.6	Cermak & Rybach 1982
Gypsum	monoclinic	2.6	3.7	1.4	Cermak & Rybach 1982
Calcite,					
0°C	trigonal	3.5	4.0	1.1	Birch & Clark 1940
50°C		3.0	3.4	1.1	Birch & Clark 1940
		3.2	3.7	1.2	Cermak & Rybach 1982
Dolomite	trigonal	4.7	4.3	0.9	Cermak & Rybach 1982
Quartz,					
0°C	trigonal	6.8	11.4	1.7	Birch & Clark 1940
50°C		5.6	9.4	1.7	Birch & Clark 1940
		6.5	11.3	1.7	Cermak & Rybach 1982
Anhydrite	orthorhombic	5.6	5.9	1.1	Cermak & Rybach 1982

some extent affect the thermal conductivity (Andrews-Speed *et al.* 1984; Brigaud & Vasseur 1989). As the grain size is found to influence the thermal conductivity, the pore size is assumed to have a corresponding effect. In fact, the measured effect of grain size shown in Fig. 5 might be a combined grain and pore size effect, as the pore size will decrease with decreasing grain size in these artificial quartz samples. How the specific pore size factor will affect the thermal conductivity and whether an increase in the mean pore size will decrease or increase the rock thermal conductivity, is not known.

There is a lack of studies that correlate thermal conductivity with the texture of sedimentary rocks. One exception is that by Penner (1963), who measured higher perpendicular thermal conductivity but lower parallel thermal conductivity for marine clays than for lacustrine clays. He concluded after thorough investigations of

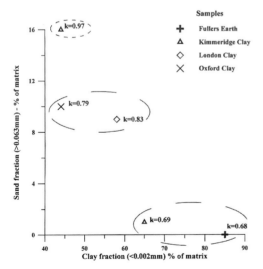

Fig. 4. Clay fraction (<2 μm) plotted v. sand fraction (>63 μm) for claystone and mudstone samples from England. The measured perpendicular thermal conductivities are shown (Midttømme *et al.* 1998a).

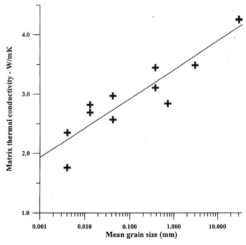

Fig. 5. Matrix conductivity, estimated by the geometric mean model, plotted against the mean grain size of artificial quartz samples. The continuous line shows the logarithmic regression, $R^2 = 0.86$ (Midttømme & Roaldset 1998).

the clays that the measured variations were due to textural differences reflecting depositional environments.

Temperature and pressure

The effects of temperature and pressure on thermal conductivity are ambiguous. Temperature is known to affect thermal conductivity, but how and to what degree is still uncertain, as both increases (Anand et al. 1973; Morin & Silva 1984; Gilliam & Morgan 1987; Somerton 1992; Midttømme et al. 1998b) and decreases (Balling et al. 1981; Brigaud et al. 1990; Demongodin et al. 1991) in thermal conductivities of clays and shales have been observed. The thermal conductivity of water increases with rising temperature, whereas the thermal conductivities of most minerals decrease with rising temperature (Clark 1966; Balling et al. 1981; Demongodin et al. 1993). The thermal conductivity of water-saturated rocks is therefore assumed to vary widely, depending upon the proportion of water to the solid part of the matrix.

An increase in measured thermal conductivity with increasing pressure has been observed (e.g. Anand et al. 1973; Gilliam & Morgan 1987) and is assumed to be due to better grain-to-grain contacts at higher pressures.

One reason for the uncertainties related to the temperature and pressure effects on thermal conductivity might be how the determination methods take into account the volume changes of the samples. Changes in pressure and temperature will induce volume changes of the pore fluid and the matrix. How the methods restrict these volume changes is important for the measured result, as heat is mainly transferred by grain-to-grain contacts. A major problem with measurements at high temperature or high pressure is the drying of the samples.

Models

Many models have been developed to estimate thermal conductivities from other known variables. However, a universal model for the thermal conductivity of sedimentary rock has not yet been formulated. The models can be grouped into three types (Somerton, 1992): (1) mixing-law models; (2) empirical models; (3) theoretical models. Mixing-law models combine values of the thermal conductivity of the rock matrix (k_s) with the conductivity of the pore fluid (k_f) on the basis of porosity. These models are of a general character and can be used for all sediment types. Thermal conductivity is related to measured physical parameters, to log data and to laboratory data in empirical models. These methods have their shortcomings in that the resulting models may be applicable only to the particular suite of rocks being investigated (Somerton 1992). Theoretical models are based on heat transfer theory for simplified geometries. The difficulty with these models is the degree of simplification necessary to obtain a solution. There is still a lack of detailed knowledge of how the heat is transferred through sedimentary rocks, and in particular, at the transition surfaces and interfaces at the grain-pore–pore-fluid and grain–grain interfaces. Preferably, one would use a theoretical model to describe the physics of heat conduction, but sufficiently reliable models have not yet been developed, and empirical modifications of the equations are needed (Zimmerman 1989; Somerton 1992).

The mixing-law models have dominated in recent thermal conductivity studies. An overview of these models has been given by, among others, Johansen (1975), Somerton (1992) and Midttømme et al. (1994). The three most commonly used, basic mixing-law models are the arithmetic mean (equation (3)), geometric mean (equation (2)) and harmonic mean (equation (4)):

$$k = \phi k_f + (1 - \phi)k_s \qquad (3)$$

$$\frac{1}{k} = \frac{\phi}{k_f} + \frac{(1 - \phi)}{k_s}. \qquad (4)$$

The harmonic and arithmetic mean models are based on parallel and series arrangements of the components relative to the direction of heat flow (Fig. 6). These two models have been used, among others, by Vacquier et al. (1988), Somerton (1992), Pribnow & Umsonst (1993) and McKenna et al. (1996). The values estimated by these models are assumed to give the upper (k_{max}) and lower (k_{min}) limits of thermal conductivity for a rock of given composition (Woodside & Messmer 1961; Zimmerman 1989; Schön 1996). In fact, they are suggested to be overpessimistic because the rock cannot be considered to be identical with the models shown in Fig. 6 (e.g. Johansen et al. 1975). Hashin & Strickman (1962) therefore modified the boundary models of the mixing-law models. The geometric mean model gives an intermediate value of the arithmetic and harmonic means (Fig. 7). This model has been widely used by, among others, Woodside & Messmer (1961), Sass et al. (1971), Balling et al. (1981), Brigaud et al. (1990), Demongodin et al. (1991), McKenna et al. (1996) and Midttømme et al. (1998b). In our opinion, the geometric mean model is the

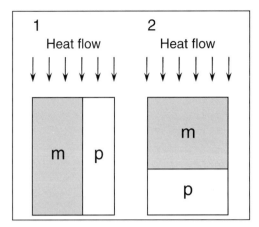

Fig. 6. Sheet models of the two mixing law models: (a) Arithmetic mean model, $k_{arith} = k_{\parallel}$; (b) harmonic mean model, $k_{harm} = k_{\perp}$. m, matrix; p, pore fluid (Schön 1996).

most realistic of these three models, as it is the only one that honours the fact that there are grain-to-grain contact paths through the material in the direction of lowest thermal conductivity.

The main criticism of the geometric mean model is that it does not take into account the texture of the samples, and is therefore valid only for isotropic rocks (Brigaud & Vasseur 1989; Somerton 1992). The harmonic, arithmetic and geometric mean models have been plotted as a function of water content in Fig. 7, together with five measurements of parallel and perpendicular

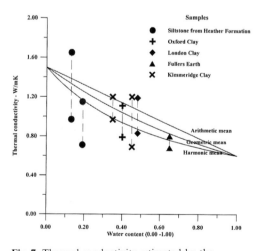

Fig. 7. Thermal conductivity estimated by the arithmetic, geometric and harmonic mean model v. water content. $k_f = 0.60$ W m^{-1} K^{-1}, $k_s = 1.50$ W m^{-1} K^{-1}. Parallel and perpendicular measured thermal conductivities of five samples are shown.

thermal conductivities. Even for these claystones and siltstones, the measured anisotropies are, with the exception of Fuller's Earth, higher than the theoretically maximum values based on arithmetic and harmonic means ($a_{max} = k_{aritm}/k_{harm}$). An explanation for this underestimate is that the arithmetic and harmonic mean models take into account only the structural arrangement of the pores and the matrix within the rocks and not the crystal anisotropy of the individual rock-forming minerals (Table 5), which is most important for the anisotropy effect in clay and mudstones (Demongodin et al. 1993). To consider this crystal anisotropy effect, a higher parallel matrix conductivity ($k_{s\parallel}$) than perpendicular matrix conductivity ($k_{s\perp}$) has to be used. This effect can also be considered by using the geometric mean model. By using the arithmetic and harmonic mean models, the estimated anisotropy depends on the arrangement of the pores and matrix. As shown in Fig. 8, the anisotropy estimated by the ratio of arithmetic and harmonic mean depends on the value of matrix conductivity (k_s), a low matrix conductivity giving low values of anisotropy. The opposite is often measured, where shales and claystones with low matrix conductivity appear to have the highest anisotropies.

The uncertainty related to the choice of the three mixing-law models depends on the ratio k_s/k_f and porosity. For water-saturated samples, where the fluid conductivity is assumed constant, the k_s/k_f ratio is determined by the matrix conductivity. According to Woodside & Messmer (1961), Lovell (1985) and Ungerer et al. (1990), the geometric mean is valid only if the k_s/k_f ratio is not too large. For ratios >20, the geometric mean model tends to overestimate substantially the measured values. For this reason, the model was shown to be appropriate for clays with low matrix conductivity (Morin & Silvia 1984). Thermal conductivities are plotted with the arithmetic, geometric and harmonic mean models as a function of the porosity in Fig. 8. The thermal conductivity of the fluid is set equal to water conductivity ($k_f = 0.60$ W m^{-1} K^{-1}) and the matrix conductivity (k_s) is constant, equal to: (a) 1.0 W m^{-1} K^{-1}, (b) 2.5 W m^{-1} K^{-1} or (c) 5.0 W m^{-1} K^{-1}. The deviation between the geometric mean model and the two other mixing-law models is shown for porosities of 10%, 20%, 40% and 60%, respectively. This discrepancy in estimated thermal conductivity varies from 1% to 68%. The highest sensitivity of the choice of mixing-law models is for the highest matrix conductivity in the porosity range of 40–60%.

The matrix conductivity (k_s) is a point of uncertainty in mixing-law models. This variable

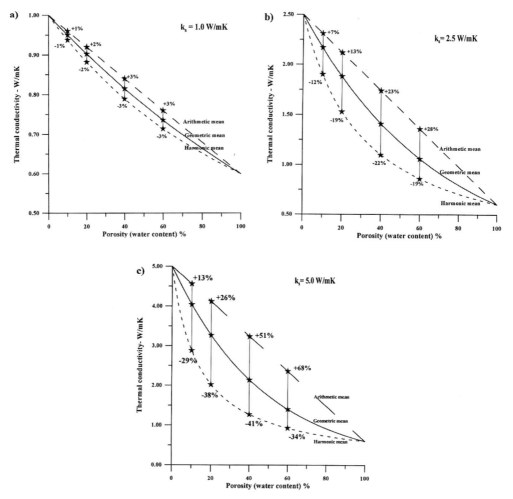

Fig. 8. Thermal conductivity estimated by the arithmetic, geometric and harmonic mean model v. porosity where $k_f = 0.60$ W m^{-1} K^{-1}, and $k_s =$ (**a**) 1.0 W m^{-1} K^{-1}, (**b**) 2.5 W m^{-1} K^{-1}, and (**c**) 5.0 W m^{-1} K^{-1}. The discrepancies between the geometric mean model and the other two mixing law models for porosities of 10%, 20%, 40% and 60% are shown.

has to take account of the mineralogical and textural effects on the thermal conductivity. Different values and models of matrix conductivity (k_s) have been used. In the simplest geometric mean models, matrix conductivity is taken to be constant (Table 6). According to Andrews-Speed et al. (1984), the matrix conductivity for real rocks varies by one order of magnitude from 0.8 to 8.3 W m^{-1} K^{-1}. Matrix conductivity is estimated from the mineralogy (e.g. Brigaud & Vasseur 1989; McKenna et al. 1996) and by use of mineralogical and grain-size data (Midttømme et al. 1998b) (Table 7) in complex models.

Although the accuracy of the estimate of the matrix conductivity depends on mineralogical and grain-size data, the right choice of model is of even greater importance. Disagreements between estimated and measured thermal conductivities are shown for claystones and mudstones in Fig. 9 and for unconsolidated sediments in Fig. 10. The thermal conductivities in these studies were estimated by the geometric mean model, where matrix conductivities were estimated from the mineralogy (Table 7). Under the assumptions that the measurements in these studies are correct, the model fails to predict the thermal conductivity for these samples.

Heat is suggested to be transferred differently through different sediment types. Somerton (1992) assumed a basic difference in the thermal characteristics of consolidated rocks and uncon-

Table 6. Thermal conductivities used in the geometric mean equation

Sample	Matrix conductivity (k_s) (W m^{-1} K^{-1})	Water conductivity (k_f) (W m^{-1} K^{-1})	Reference
Shale	2.35	0.60	Sekiguchi 1984
	1.9		Grigo et al. 1993
Sandy shale	2.1		Grigo et al. 1993
Clay, claystone, shale	3.43	0.46	Balling et al. 1981
Clay	1.2–1.4	0.60	Demongodin et al. 1991
	1.1		Grigo et al. 1993
Claystone	1.5–3	0.56	Chapman et al. 1984
Siltstone	3.2		Grigo et al. 1993
Mudstone, siltstone	2.5	0.61	Bloomer 1981
Sandy siltstone	2.49	0.60	Sekiguchi 1984
Sandy mudstone	3.0	0.61	Bloomer 1981
Shaly sandstone	2.66	0.60	Sekiguchi 1984
Muddy sandstone	3.2	0.61	Bloomer 1981
Sandstone	4.88	0.69	Balling et al. 1981
	6.60	0.60	Demongodin et al. 1991
	5–7	0.56	Chapman et al. 1984
	3.4		Grigo et al. 1993
Quartzose sandstone	7.96	0.60	Sekiguchi 1984
Quartz sandstone	3.7	0.61	Bloomer 1981
Carbonate	3.24	0.54	Balling et al. 1981
	3.0	0.59	Matsuda & Herzen 1986
Limestone	3.2	0.61	Bloomer 1981
	3.2	0.60	Demongodin et al. 1991
	3.6		Grigo et al. 1993
167 wells North Sea	0.8–8.3	0.60	Andrews-Speed et al. 1984
Gulf of Mexico	2.02	0.63	Sharp & Domenico 1976
Sedimentary rock	2.51	0.58	Smith & Chapman 1983
	1.7–4.2	0.4	Lerche 1993

solidated sand. He therefore considered modelling the two systems separately. Midttømme et al. (1997) assumed that, in a clay matrix, isolated quartz grains contribute less to the rock conductivity than quartz grains in contact with another. A distinction between matrix-supported and grain-supported sediments in modelling the thermal conductivity is also suggested. Because

Table 7. Advanced models of matrix conductivity from mineral and grain-size conductivities

Variable	Samples	Horai (1971) Pulverized samples	Brigaud & Vasseur (1989) Sandstones and artificially recompacted clay samples	Midttømme et al. (1997a) Argillaceous sediments from North Sea	Midttømme et al. (1997b) Unconsolidated sediments from Vøring Basin	
					$R^2 = 0.864$	$R^2 = 0.882$
k_f			0.60	0.60	0.60	0.60
	Quartz	7.8	7.70 ± 0.88	1.01	5.03	6.82
	Feldspar	2.3		1.02	2.97	3.49
	Calcite	3.4	3.26 ± 0.23		1.68	3.62
	Dolomite	5.1	5.33 ± 0.26		1.64	
	Pyrite	19.2		1.41		
k_s	Kaolinite	2.8	2.64 ± 0.20	0.91	0.08	2.80
	Chlorite	5.1	3.26 ± 0.25			
	Smectite		1.88 ± 0.13	1.42	3.34	1.61
	Illite	1.8	1.85 ± 0.23	1.42	3.34	1.61
	> 63 μm					5.79
	2–63 μm					1.52
	<2 μm					1.89

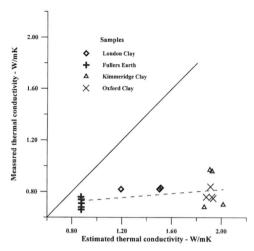

Fig. 9. Estimated thermal conductivity v. measured thermal conductivity for clay and mudstones from England. The continuous line is the unity line (1:1). The dotted line is the regression line (Midttømme et al. 1998a).

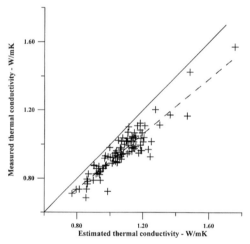

Fig. 10. Estimated thermal conductivity v. measured thermal conductivity for unconsolidated sediments from the Vøring Basin. The continuous line is the unity line (1:1). The dotted line is the regression line (Midttømme et al. 1998b).

of these distinct differences in heat transfer through sedimentary rocks, it is assumed to be impossible to develop a universal model of thermal conductivities. A number of realistic thermal conductivity models exist today, but no one specific model would be applicable to all sediment types. All models seem to be restricted either to specific rock types, ranges of variables or locations. By using the mixing-law models the restrictions are primarily due to the determination of matrix conductivity. To model these variations we suggest the development of specific matrix conductivity models for the different types of matrix. For some types of sediments, a constant matrix conductivity might serve the purpose. For other sediment types, the mineralogy, depositional environment, grain size distribution, etc., seem to affect the matrix conductivity and must be taken account in the thermal conductivity model.

Conclusions

Thermal conductivity is an important factor controlling temperature in sedimentary basins, but there are still uncertainties related to its determination. Large discrepancies in thermal conductivity measurements are observed and this warrants more basic studies, especially validation of the method of measurement for each sediment type. Standardization of the procedures of sample preparation and experimental conditions is needed to prevent unrealistic measurements.

No universal model of thermal conductivity exists today, and it will be difficult to develop such a model, as there seem to be basic differences in the mode of heat transfer through the different sediment types. Models based on the geometric mean model seem to be the most successful so far. The effect of porosity on thermal conductivity can be taken into account by this model. The matrix conductivity (k_s) in the model depends on the mineralogy and texture of the sedimentary rocks. Our experience with clays and mudstones shows that the textural influence on thermal conductivity is underestimated by many of the existing models. How the different mineralogical and textural factors will affect thermal conductivity depends on the sediment type, lithology and texture. Defining the range of application for the models according to type and texture of sediment is important for the realistic estimation of thermal conductivities for use in basin modelling.

This study has been financially supported by the Norwegian Research Council (NFR) 440.91/049 and Statoil. We gratefully acknowledge S. Lippard and F. M. Vokes, NTNU, for improving the English, and K. Gallagher, D. Grunberger and A. C. Aplin for valuable comments.

References

ANAND, J., SOMERTON, W. H. & GOMAA, E. 1973. Predicting thermal conductivities of formations from other known properties. *Society of Petroleum Engineers Journal*, 267–273.

ANDREWS-SPEED, C. P., OXBURGH, E. R. & COOPER, B. A. 1984. Temperatures and depth-dependent heat flow in western North Sea. *Bulletin, American Association of Petroleum Geologists*, **68**, 1764–1781.

BALLING, N., KRISTIANSEN, J. I., BREINER, N., POULSEN, K. D., RASMUSSEN, R. & SAXOV, S. 1981. *Geothermal measurements and subsurface temperature modelling in Denmark*. Department of Geology, Aarhus University, Geoskrifter, **16**.

BECK, A. E. 1988. Methods for determining thermal conductivity and thermal diffusivity. *In*: HAENEL, R., RYBACH, L. & STEGENA, L. (eds) *Handbook of Terrestrial Heat Flow Density Determination*. Kluwer, Dordrecht, 87–124.

BENFIELD, A. E. 1947. A heat flow value for a well in California. *American Journal of Science*, **245**, 1–18.

BIRCH, F. 1954. Thermal conductivity, climatic variation and heat flow near Calumet, Michigan. *American Journal of Science*, **252**, 1–25.

—— & CLARK, H. 1940. The thermal conductivity of rocks. *American Journal of Science*, **238**, 529–558.

BLACKWELL, D. D. & STEELE, J. L. 1989. Thermal conductivity of sedimentary rocks: measurements and significance. *In*: NAESER, N. D. & MCCULLOH, T. H. (eds) *Thermal History of Sedimentary Basins, Methods and Case Histories*. Springer, New York, 13–36.

BLOOMER, J. R. 1981. Thermal conductivities of mudrocks in the United Kingdom. *Quarterly Journal of Engineering Geology*, **14**, 357–362.

BRIGAUD, F. & VASSEUR, G. 1989. Mineralogy, porosity and fluid control on thermal conductivity of sedimentary rocks. *Geophysical Journal*, **98**, 525–542.

——, CHAPMAN, D. S. & LE DOUARAN, S. 1990. Estimating thermal conductivity in sedimentary basins using lithological data and geophysical well logs. *Bulletin, American Association of Petroleum Geologists*, **74**, 1459–1477.

CERMAK, V. & RYBACH, L. 1982. Thermal properties. *In*: HELLWEGE, K.-H. (ed.) *Landolt-Börnstein Numerical Data and Functional Relationships in Science and Technology, New Series, Group V. Geophysics and Space Research, 1*. Springer, Berlin.

CHAPMAN, D. S., KEHO, T. H., BAUER, M. S. & PICARD, M. D. 1984. Heat flow in the Uinta Basin determined from bottom hole temperature (BHT) data. *Geophysics*, **49**, 453–466.

CLARK, S. P., JR 1966. *Handbook of Physical Constants*. Geological Society of America, Memoir, **97**.

DEMONGODIN, L., PINOTEAU, B., VASSEUR, G. & GABLE, R. 1991. Thermal conductivity and well logs: a case study in the Paris Basin. *Geophysical Journal International*, **105**, 675–691.

——, VASSEUR, G. & BRIGAUD, F. 1993. Anisotropy of thermal conductivity in clayey formations. *In*: DORE, A. G., AUGUSTON, J. H., HERMANRUD, C., STEWART, D. S. & SYLTA, Ø. (eds) *Basin Modelling: Advances and Applications*. Norwegian Petroleum Society (NPF) Special Publication, **3**, 209–217.

FAROUKI, O. T. 1981. *Thermal Properties of Soils*. Cold Region Research and Engineering Laboratory Monograph **81-1**.

FJELDSKAAR, W., MYKKELTVEIT, J., CHRISTIE, O. H. J., *et al*. 1990. Interactive 2D basin modelling on workstations. *Proceedings of the SPE 20350 Petroleum Computer Conference, Denver, CO*, 181–196.

GALLAGHER, K., RAMSDALE, M., LONERGAN, L. & MORROW, D. 1997. The role of thermal conductivity measurements in modelling thermal histories in sedimentary basins. *Marine and Petroleum Geology*, **14**, 201–214.

GILLIAM, T. M. & MORGAN, I. L. 1987. *Shale: Measurement of Thermal Properties*. Oak Ridge National Laboratory ORNL/TM-10499.

GOLDSMID, H. J. & BOWLEY, A. E. 1960. Thermal conduction in mica along the planes of cleavage. *Nature*, **187**, 864–865.

GRIFFITHS, C. M., BRERETON, N. R., BEAUSILLON, R. & CASTILLO, D. 1992. Thermal conductivity prediction from petrophysical data: a case study. *In*: HURST, A., GRIFFITHS, C. M. & WORTHINGTON, P. F. (eds) *Geological Applications of Wireline Logs II*. Geological Society, London, Special Publication, **65**, 299–315.

GRIGO, D., MARAGNA, B., ARIENTI, M. T., *et al*. 1993. Issues in 3D sedimentary basin modelling and application to Haltenbanken, offshore Norway. *In*: DORE, A. G., AUGUSTON, J. H., HERMANRUD, C., STEWART, D. S. & SYLTA, Ø. (eds) *Basin Modelling: Advances and Applications*. Norwegian Petroleum Society (NPF) Special Publication, **3**, 455–468.

HASHIN, Z. & STRICKMAN, S. 1962. A variational approach to the theory of the effective magnetic permeability of multiphase materials. *Journal of Applied Physics*, **33**, 3125–3131.

HORAI, K. I. 1971. Thermal conductivity of rock-forming minerals. *Journal of Geophysical Research*, **76**, 1278–1308.

JENSEN, R. P. & DORE, A. G. 1993. A recent Norwegian Shelf heating event — fact or fantasy? *In*: DORE, A. G., AUGUSTON, J. H., HERMANRUD, C., STEWART, D. S. & SYLTA, Ø. (eds) *Basin Modelling: Advances and Applications*. Norwegian Petroleum Society (NPF) Special Publication, **3**, 85–106.

JESSOP, A. M. 1990. *Thermal Geophysics*. Elsevier, Amsterdam.

JOHANSEN, Ø. 1975. *Varmeledningevne av jordarter*. Dr.ing avhandling, institutt for kjøleteknikk, Norwegian Institute of Technology (NTH), Trondheim. (*Thermal conductivity of soils*. PhD thesis, Cold Region Research and Engineering Laboratory (CRREL) Draft Translation 637, 1977, ADA 044002, Hanover, NH.

KAPPELMEYER, O. & HAENEL, R. 1974. *Geothermics with Special Reference to Application*. Geopublication Associates, Geoexploration Monographs, Series 1, **4**.

LERCHE, I. 1993. Theoretical aspects of problems in basin modelling. *In*: DORE, A. G., AUGUSTON, J. H., HERMANRUD, C., STEWART, D. S. & SYLTA, Ø. (eds) *Basin Modelling: Advances and Appli-*

cations. Norwegian Petroleum Society (NPF) Special Publication, **3**, 35–65.

LOVELL, M. A. 1985. Thermal conductivities of marine sediments. *Quarterly Journal of Engineering Geology*, **18**, 437–441.

MATSUDA, J.-I. & VON HERZEN, R. P. 1986. Thermal conductivity variation in a deep-sea sediment core and its relation to H_2O, Ca and Si content. *Deep-Sea Research*, **33**, 165–175.

MCKENNA, T. E., SHARP, J. M., JR & LYNCH, F. L. 1996. Thermal conductivity of Wilcox and Frio sandstones in south Texas (Gulf of Mexico Basin). *Bulletin, American Association of Petroleum Geologists*, **80**, 1203–1215.

MIDDLETON, M. 1994. Determination of matrix thermal conductivity from dry drill cuttings. *Bulletin, American Association of Petroleum Geologists*, **78**, 1790–1799.

MIDTTØMME, K. & ROALDSET, E. 1997. The influence of clay minerals on thermal conductivity of sedimentary rocks. *Nordiska föreningen för lerforskning, Meddelande*, **11**.

—— & —— 1998. The effect of grain size on thermal conductivity of quartz sands and silts. *Petroleum Geoscience*, **4**, 165–172.

——, —— & AAGAARD, P. 1994. *Termiske egenskaper i sedimentoere bergarter*. Rapport fra institutt for geologi og bergteknikk, Norwegian University of Technology (NTH), **29**.

——, —— & —— 1997. Thermal conductivity of argillaceous sediments. *In*: MCCANN, D. M., EDDLESTON, M., FENNING, P. J. & REEVES, G. M. (eds) *Modern Geophysics in Engineering Geology*. Geological Society Engineering Special Publication, **12**, 355–363.

——, —— & —— 1998a. Thermal conductivity of selected clay- and mudstones from England. *Clay Minerals*, **33**, 131–145.

——, —— & BRANTJES, J. G. 1996. Thermal conductivity of alluvial sediments from the Ness Formation, Oseberg Area, North Sea. *EAEG 58th Conference, Extended Abstracts Volume 2*, P552.

——, SAETTEM, J. & ROALDSET, E. 1998b. Thermal conductivity of unconsolidated marine sediments from Vøring Basin, Norwegian Sea, Nordic Petroleum Technology Series II, 145–197.

MORIN, R. & SILVA, A. J. 1984. The effects of high pressure and high temperature on some physical properties of ocean sediments. *Journal of Geophysical Research*, **89**(B1), 511–526.

PENNER, E. 1963. Anisotropic thermal conduction in clay sediments. *Proceedings, International Clay Conference, Stockholm*. Pergamon, New York, **1**, 365–376.

PRIBNOW, D. & UMSONST, T. 1993. Estimation of thermal conductivity from the mineral composition: influence of fabric and anisotropy. *Geophysical Research Letters*, **20**, 2199–2202.

RZHEVSKY, V. & NOVIK, G. 1971. *The Physics of Rocks*. Mir, Moscow (translated by Chatterjee, A. K.).

SASS, J. H., LACHENBRUCH, A. H. & MUNROE, R. J. 1971. Thermal conductivity of rocks from measurements on fragments and its application to heat-flow determinations. *Journal of Geophysical Research*, **76**, 3391–3401.

SCHÖN, J. H. 1996. *Physical Properties of Rocks: Fundamentals and Principles of Petrophysics*. Seismic Exploration, **18**. Elsevier, Oxford.

SEKIGUCHI, K. 1984. A method for determining terrestrial heat flow in oil basinal areas. *Tectonophysics*, **103**, 67–79.

SHARP, J. M., JR & DOMENICO, P. A. 1976. Energy transport in thick sequences of compacting sediment. *Geological Society of American Bulletin*, **87**, 390–400.

SLUSARCHUK, W. A. & FOULGER, P. H. 1973. *Development and calibration of a thermal conductivity probe apparatus for use in the field and laboratory*. National Research Council of Canada, Division of Building Research, Technical Paper, **388**.

SMITH, L. & CHAPMAN, D. S. 1983. On the thermal effects of groundwater flow. 1. Regional scale systems. *Journal of Geophysical Research*, **88**(B1), 593–608.

SOMERTON, W. H. 1992. *Thermal Properties and Temperature Related Behavior of Rock/Fluid Systems*. Developments in Petroleum Sciences, **37**. Elsevier, Amsterdam.

UNGERER, P., BURRUS, J., DOLIGEZ, B., CHENET, P. Y. & BESSIS, F. 1990. Basin evaluation by integrated two-dimensional modeling of heat transfer, fluid flow, hydrocarbon generation and migration. *Bulletin, American Association of Petroleum Geologists*, **74**, 309–335.

VACQUIER, V., MATHIEU, Y., LEGENDRE, E. & BLONDIN, E. 1988. Experiment on estimating thermal conductivity of sedimentary rocks from oil well logging. *Bulletin, American Association of Petroleum Geologists*, **72**, 758–764.

VON HERZEN, R. P. & MAXWELL, A. E. 1959. The measurement of thermal conductivity of deep sea sediments by a needle-probe method. *Journal of Geophysical Research*, **64**, 1557–1563.

WOODSIDE, W. & MESSMER, J. H. 1961. Thermal conductivity of porous media. *Journal of Applied Physics*, **32**, 1688–1706.

ZIMMERMAN, R. W. 1989. Thermal conductivity of fluid-saturated rocks. *Journal of Petroleum Science and Engineering*, **3**, 219–227.

Failure envelopes of mudrocks at high confining pressures

D. N. PETLEY

Department of Geology, University of Portsmouth, Burnaby Building, Portsmouth PO1 3QL, UK

Abstract: Whereas models for the mechanical behaviour of many types of sedimentary rocks at high mean effective stresses (2–50 MPa) are now reasonably well developed, few attempts have been made to generate models for mudrocks in the same way. This paper combines a new set of undrained shear experiments on London Clay with a summary of other mudrock datasets as described in the literature, including Kimmeridge Clay, London Clay, Boom Clay and clay from mud volcanoes. It has been demonstrated previously that at low mean effective stresses, most undisturbed mudrocks behave in a brittle manner, showing a distinct peak strength before undergoing failure and strain weakening to a residual strength. At very high mean effective stresses they tend to behave in a ductile manner, with the maintenance of peak strength to large strains. This paper shows for the first time that a 'transitional' regime can be defined for most mudrocks. In this transitional regime, undrained shear deformation leads to the maintenance of peak strength to a given axial strain in a manner that is similar to ductile deformation, before the initiation of strain weakening to a residual strength. Detailed analysis has demonstrated that during the maintenance of peak strength the sample is undergoing pervasive micro-fracturing and is thus behaving in a ductile manner on the macro-scale and in a brittle manner on the micro-scale. As a result of this study it has been possible to construct a model for the form of the failure envelopes of most mudrocks, with two fundamentally important envelopes being defined: (1) a linear residual strength envelope: at low and medium mean effective stresses the stress path will tend towards this envelope after brittle failure; (2) a peak strength envelope: at low mean effective stresses this envelope has a linear form, but once the transitional regime is attained it reduces in gradient, eventually intercepting the residual strength envelope at the boundary of the ductile regime. This new framework for the behaviour of mudrocks during undrained shear deformation is generally consistent with those developed for other sedimentary rocks.

Mudrocks constitute the most abundant single rock-type in the uppermost 5 km of the Earth's crust, and a detailed understanding of the rheology of these materials is important if the mechanics of the upper part of the crust are to be fully understood. The behaviour of mudrocks is important to the hydrocarbon industry, as mudrocks often form the cap-rock of oil and gas reservoirs. Two main areas of interest have been identified. First, instability of a borehole may occur as a result of high over-pressures and/or residual tectonic stresses (Mody & Hale 1993; Abousleiman *et al.* 1995); it has been demonstrated that more than 80% of well-bore instabilities occur in mudrocks (Karfakis & Akram 1993). Such instabilities are especially common in uncased boreholes and may cause considerable difficulties. Second, the behaviour of mudrocks may be important when considering subsidence bowls above hydrocarbon reservoirs (e.g. Leddra 1989; Jones *et al.* 1992). The extraction of fluid from a hydrocarbon reservoir, and the associated decrease in local pore pressures, may induce compaction of the reservoir rocks (Jones *et al.* 1992). Displacement at the sea floor will occur if there is a transmission of strain through the overburden rocks, which will depend upon their mechanical properties. Finally, an understanding of the deformation processes in mudrocks may also have important implications for the study of deep-seated landslides (Petley 1994, 1996; Petley & Allison 1997) and in structural geology (Petley 1994).

However, the quantity of data on mudrock deformation at effective stresses above 1000 kPa is surprisingly limited. There are a multitude of data at low effective stresses (e.g. Atkinson & Bransby 1978; Cripps & Taylor 1981; Elliot & Brown 1985; Muir-Wood 1990) and during compaction (e.g. Skempton 1970; Rieke & Chilingarian 1974; Farmer 1983; Burland 1990). However, there have been few attempts to evaluate systematically the behaviour of the range of mudrocks in the same way that, for example, the behaviour of chalk has been described (e.g. Leddra 1989; Leddra et al. 1991; Kageson-Loe 1993). This paper presents a review of the behaviour of mudrocks during undrained shear deformation at high mean effective stresses (1-50 MPa) in an attempt to develop an overall model for their behaviour. It is demonstrated that it is now possible to derive a model that fits most mudrocks, although some mudrocks still appear to behave in an anomalous manner. No attempt is made to review all high-pressure mudrock experiments, but an examination of many typical examples is made and a general model is proposed.

The undrained shear deformation of mudrocks

The mudrocks described in this paper are listed in Table 1, together with the main sources of data. In most cases the samples have been subjected, in triaxial cells, to isotropic or uniaxial compaction (consolidation) followed by undrained shear deformation. Three sample states have been used in these studies: (1) undisturbed, in which the samples are unweathered and are as close to the *in situ* state of the rock as possible; (2) reconstituted, in which an undisturbed sample has been allowed to swell under slight oedometric compression; (3) remoulded, in which the sample has been disaggregated and mixed with water to a moisture content above the liquid limit. Sample preparation and testing generally conform to BS1377 (British Standards Institution 1990); further details are available in the original references.

Mud volcano clays (remoulded)

Remoulded clays have long been used in soil mechanics to establish a baseline style of behaviour and to determine the ultimate deformation state (critical state) for any given material (e.g. Atkinson & Bransby 1978; Muir-Wood 1990).

Yassir (1989) tested samples of naturally remoulded clay to investigate undrained shear deformation at confining pressures of up to 50 MPa. As the samples had been naturally remoulded they were normally consolidated and had little or no interparticle bonding.

The samples behaved in an inherently ductile manner, maintaining maximum deviatoric stress (i.e. shear stress) without dramatic strain weakening for axial strains of up to 16% (Fig. 1a), although the samples compacted to higher mean effective stresses showed limited strain hardening (increases in deviatoric stress) at high axial strains. Pore pressures increased during loading until purely ductile deformation was initiated (Fig. 1b), after which they remained approximately constant, although small decreases, probably caused by sample dilation, were sometimes observed. The samples define a linear failure envelope, with the stress paths showing decreasing mean effective stress with increasing deviatoric stress until the failure envelope was reached (Fig. 1c). At failure, deformation is either at constant mean effective stress and deviatoric stress or with small increases in mean effective stress as a result of sample dilation. The linear failure envelope can be considered to be a critical state line (Yassir 1989).

Todi Clay (remoulded, reconstituted and undisturbed)

The behaviour of Todi Clay, a low-plasticity lacustrine clay of Pleistocene age, has been described by Burland (1990). Undrained deformation experiments were conducted on undisturbed, remoulded and reconstituted samples at low and high confining pressures to investigate the form of the failure envelopes. Burland (1990) concluded that three fundamental failure envelopes could be described (Fig. 2): (1) a peak strength envelope defined by brittle failure of intact, undisturbed samples; (2) an 'intrinsic strength' envelope, defined by the

Table 1. *The materials described in this study*

Material	Source	Age	State	Reference
Trinidad Clay	Trinidad	Holocene	RM	Yassir 1989
Todi Clay	Italy	Pleistocene	RM, RC, UD	Burland 1990
Kimmeridge Clay	Kimmeridge, Dorset	Jurassic	UD	Leddra et al. 1992
London Clay	Ashford, Kent	Eocene	UD	Bishop et al. 1965
London Clay	King's Cross, London	Eocene	UD	Petley 1994
North Sea Shale	Eastern North Sea	Tertiary	UD	Petley 1994
Boom Clay	Mol, Belgium	Oligocene	UD	Horseman et al. 1993
Boom Clay	Mol, Belgium	Oligocene	UD, RC	Taylor & Coop 1993

RM, Remoulded; RC, reconstituted; UD, undisturbed.

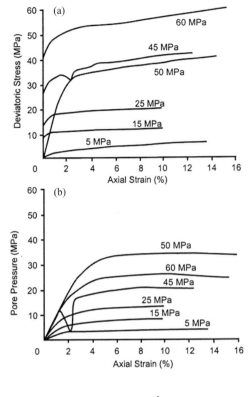

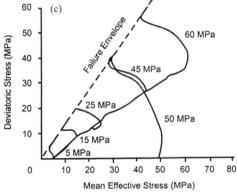

Fig. 1. Experimental data for mud volcano clays (after Yassir 1989) during undrained shear deformation. For each curve the mean effective stress before the initiation of undrained shear deformation is indicated. (**a**) Deviatoric stress–axial strain curves, illustrating the inherently ductile nature of these samples. (**b**) Excess pore pressure–axial strain curves. (**c**) Stress paths. It should be noted that many of the samples were compacted along a k_0 compaction stress path rather than the more usual isotropic stress path.

failure of reconstituted samples; this envelope was interpreted as being a basic property that is independent of the undisturbed state of the

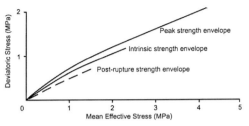

Fig. 2. The three key failure envelopes determined for Todi Clay by Burland (1990). In addition, a fourth envelope may exist: the critical state envelope with a gradient lower than that of the post-rupture strength envelope.

material, which provides a good basis for the comparison of the properties of different materials; (3) a post-rupture strength envelope, representing the end of dramatic strain weakening (i.e. reductions in deviatoric stress) in undisturbed samples after brittle failure. The envelope is determined by the amount of friction across the shear surface(s) in the sample. Further axial strain may induce gentle strain weakening to a possible critical state (residual strength) envelope.

All envelopes, with the possible exception of the critical state (residual strength) envelope, have a non-linear form, with the gradient gently decreasing with increasing mean effective stress. The differences in the slopes of the envelopes are due to differences in the fabric induced by the swelling and remoulding processes (Burland 1990).

Kimmeridge Clay (undisturbed)

Experimental investigations of the behaviour of Kimmeridge Clay, an indurated Jurassic shale from southern England, have been undertaken at high mean effective stresses by Leddra *et al.* (1991). The stress–strain curves (Fig. 3a) illustrate that all of the samples displayed brittle behaviour, although one sample (0.5 MPa) showed slight strain hardening. The two intermediate, initial mean effective stress experiments (4.5 MPa and 3.6 MPa) attain a peak strength before immediately undergoing pronounced strain weakening to a post-rupture strength (Fig. 3a). The sample deformed at the highest initial mean effective stress (10.6 MPa) also behaves in a brittle manner, although post-rupture strength is never attained, probably because the experiment was terminated prematurely at 13% axial strain (Fig. 3a). Peak strength is maintained for c. 2% axial strain before the onset of brittle failure and the mobilization of strain weakening. The stress

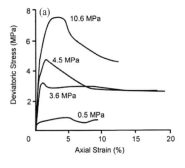

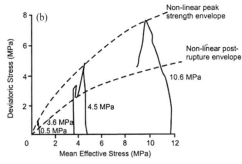

Fig. 3. Experimental data for Kimmeridge Clay (after Leddra et al. 1991) during undrained shear deformation. For each curve the mean effective stress before the initiation of undrained shear deformation is indicated. (**a**) Deviatoric stress–axial strain curves. (**b**) Stress paths, illustrating that two non-linear envelopes can be defined.

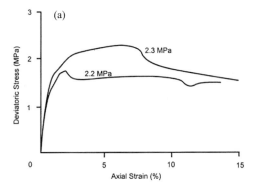

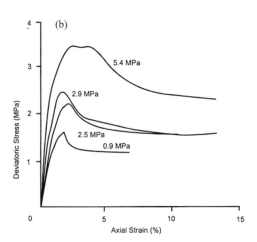

Fig. 4. Experimental data for undrained shear deformation of Boom Clay. For each curve the mean effective stress before the initiation of undrained shear deformation is indicated. (**a**) Deviatoric stress–axial strain data obtained by Taylor & Coop (1993). (**b**) Deviatoric stress–axial strain obtained by Horseman et al. (1993).

path plot shows this to be a phase of steady-state deformation on the failure envelope in a manner that is similar to ductile deformation, followed by strain weakening towards the post-rupture envelope (Fig. 3b). A non-linear peak strength envelope is defined by three of the experiments, suggesting that the result of the 4.5 MPa experiment is anomalous (Fig. 3b). A non-linear post-rupture envelope is also defined by all four samples.

Boom Clay (reconstituted and undisturbed)

The behaviour of samples of Boom Clay at mean effective stresses between 200 kPa and 5.42 MPa has been investigated by Taylor & Coop (1993) and Horseman et al. (1993). In both cases samples were collected from a shaft c. 247 m below the ground surface. Triaxial experiments were conducted on a total of six samples, four of which were undisturbed whereas the other two were reconstituted.

The experiments of Taylor & Coop (1993) were mostly conducted under extension, but two tests were undertaken under compression. The results of the experiments are somewhat anomalous (Fig. 4a). The sample deformed at an initial mean effective stress of 2.3 MPa showed the maintenance of peak strength to an axial strain of about 7% before abrupt brittle failure and strain weakening to a post-rupture strength. However, the sample deformed at 2.2 MPa underwent brittle failure at a lower deviatoric stress and at an axial strain of about 2%. According to Taylor & Coop (1993), the samples appear to define a linear brittle failure envelope with $M = 0.95$. The post-rupture envelope is also common to both experiments and has a gradient of $M = 0.85$.

The experiments of Horseman et al. (1993) are shown in Fig. 4b. Stress path plots are not available, but the stress paths traced are

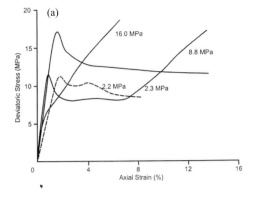

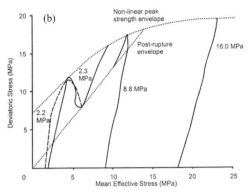

Fig. 5. Experimental data for Eocene North Sea Shale (after Petley *et al.* 1993). For each curve the mean effective stress before the initiation of undrained shear deformation is indicated. (**a**) Deviatoric stress–axial strain data. (**b**) Stress path data, illustrating the form of the failure envelopes.

described as being unusually linear, similar to those for the London Clay (Petley 1994) and the Kimmeridge Clay (Leddra *et al.* 1991). The post-rupture envelope was reported to be approximately linear with $M = 0.81$. In stress–strain space the samples are brittle with post-rupture strain weakening (Fig. 4b). However, the sample deformed at the highest initial mean effective stress (2.3 MPa), undergoes a phase of pseudo-ductile deformation at constant deviatoric stress before strain weakening, whereas the other samples do not (Fig. 4b).

Eocene North Sea Shale (undisturbed)

Undrained shear deformation experiments undertaken on four samples of undisturbed, indurated shale from the Norwegian sector of the North Sea have been described by Petley *et al.* (1993, 1994) and Petley (1994). The data suggest that the behaviour of this material is highly complex and requires considerable further research, although this is hindered by the difficulties of obtaining samples. The results display increasing mean effective stress and deviatoric stress until a peak strength is attained (Fig. 5a), defining a non-linear peak strength envelope. At peak strength the samples showed abrupt strain weakening to a linear post-rupture envelope (Fig. 5b). Two of the samples thereafter remained on this line, with the 2.2 MPa sample at a steady state and the 8.8 MPa sample showing a steady decrease in both deviatoric stress and mean effective stress (Fig. 5b). The 2.3 MPa sample remained at a steady state for about 4% axial strain (Fig. 5a) before strain hardening and defining a stress path that was steeper than the post-rupture envelope (Fig. 5b). Such behaviour is anomalous and may be associated with an equipment fault; it has not been seen in any other mudrock experiments.

London Clay

Undrained shear deformation experiments on samples of London Clay at high effective stresses have been undertaken by Bishop *et al.* (1965) and Petley (1994). The experiments of Petley (1994) sought to define the form of the peak strength (brittle failure) envelope. The samples behaved in a manner similar to that of a lightly over-consolidated or normally consolidated clay (e.g. Atkinson & Bransby 1978) (Fig. 6), showing little change in mean effective stress with increasing deviatoric stress (Fig. 6b). The samples define a non-linear peak strength envelope before undergoing strain weakening to a non-linear post-rupture envelope. At the lowest initial mean effective stress the peak strength envelope corresponds to that defined by low-pressure (100–1000 kPa) triaxial experiments (e.g. Atkinson & Bransby 1978), having an approximately linear form with a gradient of $M = 0.89$. At higher initial mean effective stresses the peak strength envelope has a notable curvature, the gradient decreasing to $M = 0.64$ at 30 MPa. All samples strain weaken after peak strength to a non-linear post-rupture, which is similar to that measured in experiments at low mean effective stresses (Schofield & Wroth 1968; Atkinson & Bransby 1978).

The stress–strain curves suggest that the samples underwent brittle failure (Fig. 6a), but peak strength is maintained before failure in a manner that is similar to ductile deformation. The axial strain at which peak strength is maintained is dependent on the magnitude of the initial mean effective stress. Thus the samples display a combination of ductile behaviour, in

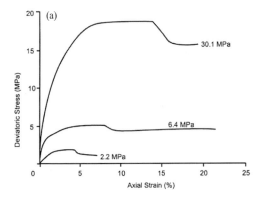

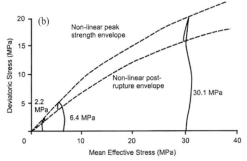

Fig. 6. Experimental data for London Clay (after Petley 1994) during undrained shear deformation. For each curve the mean effective stress before the initiation of undrained shear deformation is indicated. (**a**) Deviatoric stress–axial strain. (**b**) Stress paths and failure envelopes, illustrating the non-linear form of the failure envelopes.

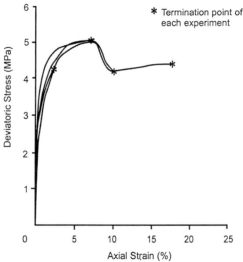

Fig. 7. Deviatoric stress–axial strain data for London Clay experiments deformed at an initial mean effective stress of 6.4 MPa (after Petley 1994). Each of the four experiments was terminated at a different axial strain as indicated by an asterisk, allowing physical examination of the sample to be undertaken.

which peak strength is maintained, and brittle failure, with pronounced strain weakening to a residual strength. This behaviour is similar to that displayed by the Kimmeridge Clay.

The nature of the ductile–brittle behaviour was investigated in a further set of London Clay experiments (Petley 1994). In each case the experiment was terminated at a predetermined axial strain (Fig. 7). Physical examination of each sample upon removal from the triaxial cell allowed insight to be gained into the deformation processes acting on the sample. During the 'ductile' phase the strains are not localized but remain distributed through the sample. Physical examination of the samples using a scanning electron microscope demonstrated that the sample is undergoing pervasive micro-cracking, such that on the micro-scale the deformation is brittle. However, on the macro-scale (whole sample) the deformation is uniformly distributed and is effectively ductile. As axial strain increases the micro-cracks grow and coalesce. Brittle failure is initiated when the micro-cracks coalesce sufficiently for a single plane of weakness to form across the sample. Thereafter, the sample behaves in a manner similar to that of a brittle material.

Petley (1994) demonstrated that pore pressures increase as deviatoric stress increases and remain approximately constant during the 'ductile' phase. If the sample is undergoing micro-cracking, the sample must also be undergoing local dilation, but the fluid connection between the micro-cracks is insufficient to allow the localized dilation to affect the rest of the sample.

The behaviour of London Clay under high effective stresses was also investigated by Bishop *et al.* (1965). These data have also been reinterpreted by Burland (1990). The form of the failure envelopes deduced by Bishop *et al.* (1965) are illustrated in Fig. 8a. Failure envelopes are plotted for intact samples from a range of depths and reconstituted samples from 34.8 m. The envelopes have a non-linear form, with the gradient decreasing at higher initial mean effective stresses. Stress–strain curves for the experiments (Fig. 8b) show that at low and medium stresses the London Clay behaves in an intrinsically brittle manner, but at moderate stresses the peak strength is maintained before strain

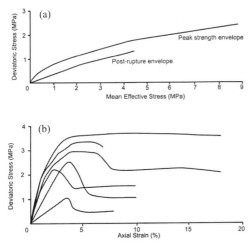

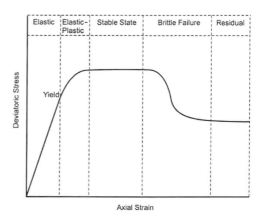

Fig. 9. Generalized form of the deviatoric stress–axial strain graph for mudrocks deforming in the brittle–ductile transition.

Fig. 8. Experimental data for London Clay (after Bishop *et al.* 1965; Burland 1990) during undrained shear deformation. For each curve the mean effective stress before the initiation of undrained shear deformation is indicated. (**a**) Failure envelopes, illustrating the non-linear form of the peak strength and post-rupture envelopes. (**b**) A family of deviatoric stress–axial strain curves.

weakening, a behaviour that is similar to that shown in the experiments of Petley (1994). Like the Todi Clay (Burland 1990) and the London Clay results of Petley (1994), the post-rupture envelope has a non-linear form, with the gradient decreasing at higher mean effective stresses.

Discussion

The data presented in the preceding sections highlight the difficulties of developing a unifying model for the deformation of all mudrocks. These are partly a result of the limited range of materials that have been tested, meaning that there is no database for the behaviour of any mudrock that approaches that which has been used to model, for example, the behaviour of chalk. Also, there is probably a great variation in the properties of mudrocks, including variables such as mineralogy, particle size and shape, stress state, void ratio, fabric, cementation and interparticle bonding. Additionally, almost all of the experimental programmes described here have individual results that do not fit the pattern shown by the rest. However, despite these variations in behaviour, some interesting general patterns of behaviour that characterize many of the materials have emerged. Thus it is now possible to propose a model that describes the behaviour of many of the mudrocks.

The brittle–ductile transition

The data collected by Bishop *et al.* (1965) and Petley (1994) for the London Clay, Horseman *et al.* (1993) and Taylor & Coop (1993) for the Boom Clay, and Leddra *et al.* (1991) for the Kimmeridge Clay, appear to fit into a similar pattern, interpreted as the brittle–ductile transition. The transition is very important in terms of changes in the rheological properties of rocks but has been rarely described for mudrocks, although it has been outlined in great detail for a diverse range of other materials, including limestone (Donath *et al.* 1964), sandstone (Rutter & Hazibadeh 1991) and chalk (Leddra 1989). Examination of the stress–strain curves for the London Clay, Boom Clay and Kimmeridge Clay shows that, in each case, samples deformed at the higher confining pressures maintain peak strength before undergoing brittle failure and strain weakening (e.g. see Fig. 7 for London Clay). Thus, deformation is a combination of ductile behaviour and brittle failure. The generalized form of this stress–strain plot has been deduced (Fig. 9). After an initial phase of elastic deformation, the sample undergoes yield and a phase of elastic–plastic deformation. Eventually, a stable state is reached in which peak strength is maintained as strain accumulates whereas pore pressures remain constant. After a given axial strain has accumulated, brittle failure occurs as a result of the formation of a single fracture caused by the coalescence of microcracks formed during the ductile deformation phase. At this point the sample undergoes strain weakening to a post-rupture state defined by

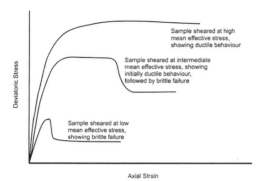

Fig. 10. Generalized family of deviatoric stress–axial strain graphs for mudrocks deforming in the brittle, transitional and ductile regimes.

friction across the shear plane. At very high axial strains, gentle strain weakening may occur until the critical state (residual strength) is attained.

Brittle failure occurs at large strains with the type of deformation being dependent upon the initial mean effective stress of the sample. At higher stresses the peak state is maintained for greater axial strains. Therefore, a general model for the behaviour of these materials through the brittle–ductile transition can be deduced (Fig. 10). At low mean effective stresses deformation is brittle in nature with a distinct peak strength, followed by strain weakening to a post-rupture value. At very high mean effective stresses the response is ductile in nature, with peak strength being maintained for the accumulation of very large strains. Between these two extremes there is a transitional regime in which deformation is a combination of ductile behaviour, in which peak strength is maintained, followed by brittle behaviour, in which failure and strain weakening occur. Such behaviour is similar to that displayed by other geological materials such as the Crown Point limestone (Donath *et al.* 1964).

Failure envelopes

Various failure envelopes have been defined for the different mudrocks described in this paper. It is proposed here that three envelopes are important in understanding the behaviour of an undisturbed mudrock, although it is recognized that an intrinsic failure envelope (Burland 1990) is important if an understanding of the full mechanical properties of an undisturbed mudrock is to be gained. A fundamental property of a mudrock is the critical state, as ultimate deformation will tend towards this line. Also, the peak strength (brittle) envelope is important, as this will define the strength properties of the material. Finally, the post-rupture envelope, to which the sample will trend after brittle failure, has important geotechnical applications (Burland 1990), such as in the understanding of deep-seated landslides (Petley 1996).

Previous studies have defined the geometries of these envelopes for various materials (e.g. Burland 1990). The experiments examined in this study allow a new model to be proposed for the form of the envelopes. In some cases the exact form of the envelope has still to be defined; clearly, further research is needed.

In all of the experiments on intact material it is possible to define a peak strength envelope that can be equated with a Mohr–Coulomb failure envelope. Many experimental programmes in the low-pressure (0–700 kPa) confining pressure regime have suggested that through most of this regime it has a linear form, although at very low pressures (<100 kPa) it may be non-linear. There is clear evidence from the current study that at higher confining pressures the form is non-linear, with the gradient steadily reducing with increasing mean effective stress. This reduction in slope appears to be coincident with the initiation of the brittle–ductile transitional regime, and it is probable that the two are linked.

The form of the post-rupture envelope is less clear. Again, at low confining pressures it has a linear form. For higher confining pressures it appears to have a linear form for Boom Clay, but the work of Bishop *et al.* (1965), Petley (1994) and Burland (1990) suggests that it also reduces in gradient with increasing mean effective stress for London Clay. This envelope may be difficult to define, as the position may be heavily dependent on strain-rate effects. Finally, there is a critical state (residual strength) envelope, which the data of Yassir (1989) suggest is linear at up to 50 MPa.

Thus it is possible to plot a conceptual diagram of the three envelopes (Fig. 11a and b). The three envelopes have a simple form. The linear critical state (residual strength) envelope is not attained except at very high strains for low confining pressures (generally <1 MPa). However, at very high effective stresses, when compaction has broken many of the interparticle bonds and the material behaves in a ductile manner, it is the main deformation envelope. At low mean effective stresses the samples can sustain stresses far greater than those predicted by the critical state line, and a linear brittle failure envelope is defined. After rupture, the samples strain weaken to the post-rupture envelope. At very large strains, it is probable that they will continue to strain weaken and will

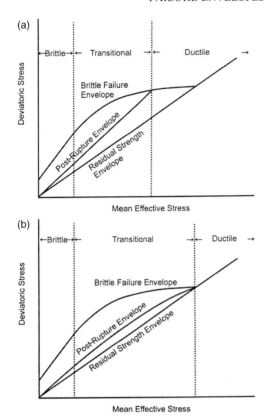

Fig. 11. Conceptualized failure envelopes for mudrocks illustrating the two different forms suggested by data in this study. (**a**) Envelopes of a material with a linear post-rupture envelope, such that it intersects the brittle failure envelope at a lower mean effective stress than the intersection of the residual strength envelope and the brittle failure envelope. (**b**) Envelopes of a material with a non-linear post-rupture envelope, such that it intersects the brittle failure envelope at the same mean effective stress as that of the residual strength envelope and the brittle failure envelope.

tend towards the linear critical state (residual strength) envelope. At moderate stresses the stress paths intersect and remain on a continuation of the brittle failure envelope, which is non-linear in this regime, before strain weakening to the post-rupture envelope. It is proposed that the non-linearity of the envelope represents the initiation of the brittle–ductile transition.

At present, the form of the envelopes is not clearly defined at the high mean effective stresses end of the transition regime. If the post-rupture envelope has a linear form, it will intersect the peak strength envelope (Fig. 11a). Therefore, a sample being deformed to the left of this point will undergo strain weakening to the post-rupture surface, although the amount of weakening will decline with increasing mean effective stress. To the right, the samples will not undergo brittle failure, possibly because the micro-cracks are too fine to facilitate this, but will enter a phase of pseudo-ductile deformation. A small amount of weakening will occur as the stress path will tend towards the critical state (residual strength) envelope. At very high mean effective stresses, the peak strength envelope coincides with the critical state (residual strength) envelope, beyond which a sample will deform in a truly ductile manner.

Alternatively, the post-rupture envelope may have a non-linear form (Fig. 11b), with the brittle failure envelope and the residual strength envelope having the same form as before. The post-rupture envelope will intersect the brittle failure envelope at the junction with the critical state (residual strength) line, meaning that a clear termination for the transitional zone is possible. At present, it is not possible to say for certain which of the two models is correct.

The basic form of the brittle failure envelope and the residual strength envelope is the same as that for chalk, although the brittle failure envelope appears to have a linear form. However, there are insufficient detailed data to confirm definitively that this is the case. More importantly, chalk does not appear to have a post-rupture envelope. This is probably attributable to the sensitive nature of the material, meaning that, as a result of the high porosity, it tends to collapse and remould at the point of failure. In consequence, it will begin to behave in a remoulded fashion and will therefore strain weaken almost instantaneously to the critical state (residual strength) line. The same process is probably responsible for the lack of a brittle–ductile transition: as soon as the structure begins to undergo micro-cracking it collapses, replicating brittle failure and preventing the transitional regime from occurring. It is only at very high mean effective stresses, when the structure of the chalk has been sufficiently densified, that purely ductile behaviour can be initiated, at which point the brittle–failure envelope and the residual strength line coincide. This is confirmed by the behaviour of the remoulded chalk samples tested by Kageson-Loe (1993), which appear to display a transitional regime. These samples have the same mineralogy as the undisturbed samples but do not have the high preserved pore volume, such that collapse cannot occur. It is, however, interesting that remoulded chalk does display transitional behaviour whereas remoulded mudrocks do not.

It is possible that many of the anomalous results seen in the various mudrock programmes are attributable to a similar problem. In many cases, these results show samples undergoing premature brittle failure and thus never reaching the peak strength envelope. It is proposed that this early brittle failure is the result of imperfections in the sample such as minor cracks or relatively large clasts that might concentrate stress. Failure may occur on or around these imperfections, inducing an early strain weakening phase.

Mudrocks that do not fit the model

The model proposed above describes the behaviour of London Clay, Boom Clay, Kimmeridge Clay and, to a lesser extent, Todi Clay. The remoulded clays of Yassir (1989) behave as a typical homogeneous, unbonded material, deforming on a true critical state line as a lack of inter-particle bonding or cement prevents brittle deformation. Similarly, the reconstituted clays of Burland (1990) do not behave as the undisturbed material, possibly because the swelling phase damages the inter-particle bonds. However, the Eocene North Sea Shale is not described by the model. This may be the result of a single influence, or a combination of a number of different influences, including the following: a different inherent property because of the state of lithification of the rock; different response because of the low porosity of the rock; the depth of burial has allowed very low-grade metamorphism to change the rock properties; damage to the structure of the rock as a result of the coring process. It is impossible to estimate which of these influences is responsible for the different material behaviour. Much more detailed research is needed to investigate the different properties of these materials, although samples are exceptionally difficult and expensive to obtain.

Conclusions

A review has been made of the undrained shear deformation of various mudrocks at high effective stresses. It has been demonstrated that most of the mudrocks show a similar response to deformation in the high-pressure range, and display a transitional regime between brittle and ductile behaviour. It has been proposed that this transitional regime leads to a decrease in the gradient of the failure envelope, as the microcracking prevents the material from sustaining high mean effective stresses. Eventually, the gradient of the envelope decreases such that it intersects the residual strength envelope; thereafter, deformation is in the ductile regime.

Samples deforming in the brittle regime will undergo strain weakening after failure to the post-rupture line; thereafter, they may undergo slow strain weakening and will, at very large strains, reach the residual critical state line (Burland 1990). The form of the post-rupture envelope is not clear at this stage. Some experiments have suggested that it is linear, in which case it will intersect the brittle failure line, whereas others suggest that, like the brittle failure envelope, it reduces in gradient and may intersect the brittle failure line at its intersection with the critical state (residual strength) line.

It is shown that in terms of the peak strength and residual strength line, mudrocks have a similar failure envelope geometry to chalk. However, chalk does not have a transitional regime or a post-rupture line, probably because it collapses and becomes remoulded upon failure.

References

ABOUSLEIMAN, Y., INDERHAUG, O., DIEK, A. L. & ROEGIERS, J.-C. 1995. Effects of fluids on the fracture toughness of shale under mode I load conditions. *Proceedings of the International Colloquium on Chalk and Shales.* GBMR and BVRM, Brussels, 2.1.1–2.1.10.

ATKINSON, J. H. & BRANSBY, P. L. 1978. *The Mechanics of Soils.* McGraw–Hill, London.

BISHOP, A. W., WEBB, D. L. & SKINNER, A. E. 1965. Triaxial tests on soil at elevated cell pressures. *Proceedings of the International Conference of Soil Mechanics, Montreal,* **1**, 170–174.

BRITISH STANDARDS INSTITUTION 1990. *BS1377 part 2. British Standard Methods of Test for Soils for Civil Engineering Purposes.* British Standards Institution.

BURLAND, J. 1990. On the compressibility and shear strength of natural clays. *Géotechnique,* **40**, 329–378.

CRIPPS, J. C. & TAYLOR, R. K. 1981. The engineering behaviour of mudrocks. *Quarterly Journal of Engineering Geology,* **14**, 325–346.

DONATH, F. A., FAILL, R. T. & TOBIN, D. G. 1971. Deformation mode fields in experimentally deformed rock. *Geological Society of America Bulletin,* **82**, 1441–1462.

ELLIOT, G. M. & BROWN, E. T. 1985. Yield of a soft, high porosity rock. *Géotechnique,* **35**(4), 413–423.

FARMER, I. 1983. *Engineering Behaviour of Rocks.* Chapman and Hall, New York.

HORSEMAN, S. T., WINTER, M. G. & ENTWISTLE, D. C. 1993. Triaxial experiments on Boom Clay. *In*: CRIPPS, J. C., COULTHARD, J. M., CULSHAW, M. G., FORSTER, A., HENCHER, S. R. & MOON, C. F. (eds) *The Engineering Geology of Weak Rock.* Balkema, Rotterdam.

JONES, M. E., LEDDRA, M. J., GOLDSMITH, A. S. & EDWARDS, D. 1992. *The Geomechanical Characteristics of Reservoirs and Reservoir Rocks.* OTH 90 333. HMSO, London.

KAGESON-LOE, N. M. 1993. *The strain behaviour of chalk*. PhD thesis, University of London.

KARFARKIS, M. G. & AKRAM, M. 1993. Effects of chemical solutions on rock fracturing. *International Journal of Rock Mechanics, Mineral Science and Geomechanics Abstracts*, **30**, 1253–1259.

LEDDRA, M. J. 1989. *Deformation of chalk through compaction and flow*. PhD thesis, University of London.

——, PETLEY, D. N. & JONES, M. E. 1991. Fabric changes induced in a cemented shale through consolidation and shear. *In*: TILLERSON, J. R. & WAWERSIK, W. R. (eds) *Rock Mechanics*. Balkema, Rotterdam, 917–926.

MODY, F. K. & HALE, A. H. 1993. Borehole-stability model to couple mechanics and chemistry of drilling fluid/shale interactions. *Journal of Petroleum Technology*, **45**, 1093–1101.

MUIR-WOOD, D. 1990. *Soil Behaviour and Critical State Soil Mechanics*. Cambridge University Press, Cambridge.

PETLEY, D. N. 1994. *The deformation of mudrocks*. PhD thesis, University of London.

—— 1996. The mechanics and landforms of deep-seated landslides. *In*: BROOKS, S. & ANDERSON, M. (eds) *Advances in Hillslope Processes*. Wiley, Chichester, 823–836.

—— & ALLISON, R. J. 1997. The mechanics of deep-seated landslides. *Earth Surface Processes and Landforms*, **22**, 747–758.

——, JONES, M. E., STAFFORD, C., LEDDRA, M. J. & KAGESON-LOE, N. L. 1993. Deformation and fabric changes in weak fine-grained rocks during high pressure consolidation and shear. *In*: ANAGNOSTOPOULIS, A., SCHLOSSER, F., KALTZIOTIS, N. & FRANK, R. (eds) *Geotechnical Engineering of Hard Soils–Soft Rocks*, 1, Balkema, Rotterdam, 371–382.

——, LEDDRA, M. J., JONES, M. E. & KAGESON-LOE, N. M. 1994. On fabric changes in a cemented material during consolidation and shear. *In*: AASEN, J. O., BERG, E., BULLER, A. T., HJELMELAND, O., HOLT, R. M., KLEPPE, J. & TORSAETER, O. (eds) *North Sea Oil and Gas Reservoirs III*. Balkema, Rotterdam, 371–382.

RIEKE, H. H. & CHILINGARIAN, G. V. 1974. *Compaction of Argillaceous Sediments*. Elsevier, Amsterdam.

RUTTER, E. H. & HADIZADEH, J. 1991. On the influence of porosity on the low-temperature brittle–ductile transition in siliciclastic rocks. *Journal of Structural Geology*, **13**, 609–614.

SCHOFIELD, A. N. & WROTH, C. P. 1968. *Critical State Soil Mechanics*. McGraw–Hill, London.

SKEMPTON, A. W. 1970. The consolidation of clays by gravitational compaction. *Quarterly Journal of Engineering Geology*, **125**, 373–411.

TAYLOR, R. N. & COOP, M. R. 1993. Stress path testing of Boom Clay from Mol, Belgium. *In*: CRIPPS, J. C., COULTHARD, J. M., CULSHAW, M. G., FORSTER, A., HENCHER, S. R. & MOON, C. F. (eds) *The Engineering Geology of Weak Rock*. Balkema, Rotterdam, 77–82.

YASSIR, N. A. 1989. *Mud volcanoes and the behaviour of overpressured clays and silts*. PhD thesis, University of London.

Principal aspects of compaction and fluid flow in mudstones

K. BJØRLYKKE

Department of Geology, University of Oslo, PO Box 1047, 0316 Oslo, Norway

Abstract: Compaction and fluid flow in sedimentary basins are usually modelled as mechanical compaction assuming that the main driving force is the effective stress. However, in the deeper part of sedimentary basins (>2–3 km, 70–100°C), compaction also involves dissolution and precipitation of minerals, and these processes are strongly controlled by temperature and to a lesser extent by variations in effective stress. The total pore-water flux is relatively independent of the permeability and pressure gradients because it is controlled by the rate of compaction. The main effect of permeability variations is the focusing of the compaction-driven flow. Fluids are also released by petroleum generation and dehydration of clay minerals; both these processes are also essentially temperature dependent. If the minimum permeability in one sedimentary layer (seal) falls below a critical value and there is little lateral drainage, the pressure will build up to fracture pressure. Pore pressures and fluid flow cannot be calculated from the matrix permeability of the rock when fracture pressures are reached, because much of the flow occurs along the generated fractures. The permeability produced by fracturing is a dynamic variable adjusting itself to the flux; it is not a rock property like the intergranular permeability. The intergranular permeability of shale samples before fracturing is also difficult to measure reliably in the laboratory as a result of unloading. This makes pressure prediction using one-dimensional pressure modelling uncertain. The prediction of permeability distributions in three dimensions is much more difficult.

Problems related to fluid flow in sedimentary basins, and in mudstones in particular, can be approached from several different directions, depending on the background of the researcher, which may be geophysics, clay mineralogy, diagenesis, soil and rock mechanics or mathematical modelling.

Compaction of mudstones and shales is mostly treated as a mechanical process, following classical soil and rock mechanical theory, and is then assumed to be a function of the effective stress (Rieke & Chilingarian 1974). This may be valid for the shallow parts of sedimentary basins, but below 2–3 km depth (70–100°C), mineral dissolution and precipitation become more important.

Most basin modellers treat compaction as a direct function of stress, even in the deepest part of the basin (Illiffe & Dawson 1996; Yu & Lerche 1996). Chemical compaction, however, follows very different 'laws', and temperature and mineralogy are the most important controls. The results of the modelling will be dramatically different depending on whether compaction is modelled as a mechanical or chemical process. The purpose of the present paper is to discuss some of the principal aspects of mud and shale compaction, which should be considered when interpreting data from rocks and fluids in sedimentary basins.

Mechanical compaction

The degree of compaction that results from a certain effective stress (compressibility) is a function of the grain size distribution and mineralogy. Compaction is a result of grain reorientation and breakage causing a reduction in porosity (ϕ) (Fig. 1). Shales and mudstones contain not only soft clay minerals, but in most cases also clay- and silt-sized quartz, feldspar and mica. The compressibility therefore varies within wide limits depending on the lithology. This can be expressed by the beta factor, which relates the change in void ratio ($\phi/(1-\phi)$) with an increase in stress (Aplin *et al.* 1995). Experimental data show that pure fine-grained clay minerals do not compact to porosity values below 25%, even when subjected to stress equivalent to 5 km overburden (Rieke & Chilingarian 1974). Pure montmorillonite does not compact more than to 45% porosity under the same conditions, and this is probably due to the large surface area of such fine clays, which also have much bound

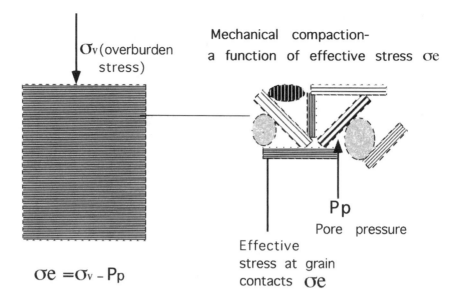

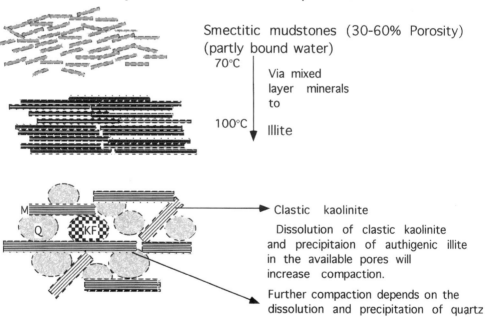

Fig. 1. Schematic representation of mechanical and chemical compaction in mudstones. M, micas; KF, K-feldspar; Q, quartz.

water. Also, sandstones compact much less by experimental mechanical compaction than is normally observed in sedimentary basins. Even sandstones with 50% quartz grains and 50% softer lithic fragments have >20% porosity after being compacted experimentally to stresses equal to almost 5000 m of overburden (Pittman & Larese 1991). This is much higher porosity than is normally observed and shows that experimental mechanical compaction cannot realistically simulate natural compaction in sedimentary basins.

Chemical compaction

Chemical compaction involves dissolution and precipitation of minerals. The kinetics of these reactions is strongly controlled by temperature, particularly for silicate minerals. Carbonate diagenesis is less temperature dependent than are silicate reactions because of higher kinetic reaction rates at low temperatures.

Chemical compaction is driven towards a higher degree of thermodynamic stability (lower free energy), and the kinetics determines the rate of the mineral reactions. Dissolution of aragonite and precipitation of calcite cause chemical compaction when aragonite fossils are parts of the grain framework. Dissolution of aragonite and precipitation of calcite cement in mudrocks drastically change the rock's properties at shallow depth.

Minerals formed by weathering at low temperature near the surface, such as smectite, gibbsite and kaolinite, will in most cases dissolve when temperatures are raised at greater burial (Bjørlykke 1998). The temperature at which these reactions occur depends, however, on a supply of cations, primarily potassium, from potassium-feldspar, so that illite can precipitate as the most stable diagenetic mineral.

A shale consisting of quartz, feldspar, mica and illite will be thermodynamically stable up to very high temperatures (greenschist or amphibolite facies) because there are no other mineral assemblages that are more stable. Quartz will gradually be redistributed by pressure solution and illite will become more crystalline. It is only in the presence of minerals such as smectite and kaolinite that K-feldspar becomes unstable during burial diagenesis.

Many of the same diagenetic processes that have been studied in sandstones also apply to mudstones, but they are more difficult to study because of the fine-grained matrix. Also, in mudstones pore space is present because the overburden stress is supported by a grain framework where the larger grains may be quartz and carbonate. They may dissolve by pressure solution and thus allow further compaction. This process may occur at shallow depth in the case of carbonate minerals, but in the case of quartz, the temperature is the rate-limiting factor, as it is in the case of sandstones (Bjørkum 1995; Walderhaug 1996; Bjørkum & Nadeau 1998). Clastic kaolinite may also be relatively coarse grained and be part of the grain framework in mudstones. Dissolution of such grains may cause significant chemical compaction because they may be parts of the load-bearing grain framework in shales.

Sediment compaction and fluid flow

Darcy's law can for this purpose be expressed as

$$F = k \times \text{Grad } P/\mu$$

where F is the compaction-driven flux as a function of the reduction in porosity of the underlying sequences, k is the permeability of the least permeable layers (seal), Grad P is the potentiometric gradient and μ is the viscosity of the fluid.

Darcy's law is often used to model fluid flow by estimating the pressure gradients and permeability of shales. However, the total fluid flux (F) is limited by the rate of compaction and porosity loss of the underlying sediments. In modelling of fluid flow in sedimentary basins, the supply of fluid must be the main constraining factor rather than the pressure gradients.

In the case of mechanical compaction, the rate of porosity reduction, and thereby the pore water flux, is related to the rate of increase in effective stress. If compaction results in a significant increase in pore pressures above hydrostatic pressure, the effective stress is reduced and compaction may be reduced or stop altogether (Fig. 2). A negative feedback therefore exists between overpressure and fluid flow. It is clear from Fig. 2 that the mechanical compaction cannot generate and sustain overpressures up to fracture pressures, because the effective stress will then remain almost constant during burial and no further mechanical compaction will occur. If the underlying sediments are much more permeable than the seal at the top of the overpressure, the effective stress will gradually increase with depth following a hydrostatic pressure gradient from the fracture pressure (Fig. 2). It is then possible that some mechanical compaction may occur in the deeper parts of the basin, but most of the sequence is probably overcompacted compared with previous maximum stress, implying no compaction-driven flow.

Highly overpressured compartments may be bounded by shales that serve as seals with high-pressure gradients (Hunt 1990). However, the pressure gradients observed cannot be used to calculate the flow across the seal because the effective permeability of the seal cannot be determined.

This is because normally only cuttings are obtained from these shales, and even samples from cores may develop fractures as a result of

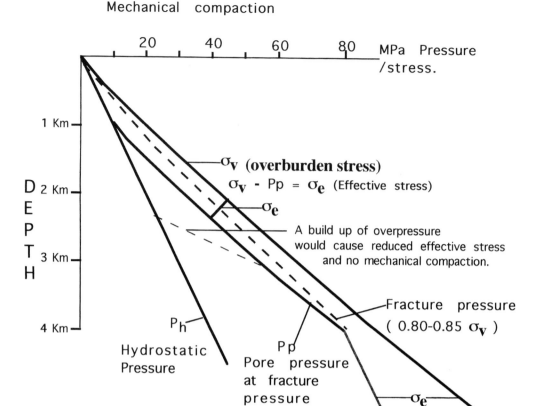

Fig. 2. Relationships between overburden stress, fracture pressure and hydrostatic pressure as a function of burial depth. Mechanical compaction is a function of effective stress, which is the difference between the overburden stress (σ_v) and the pore pressure (P_p). At 4 km the effective stress (σ_e) at fracture pressure is about 10 MPa, which is equal to the difference between the total stress (90 MPa) and the pore pressure (80 MPa) at 2.00 g cm^{-3}. This is equal to the effective stress at 1 km with normal hydrostatic pressure (P_h).

unloading that may cause unreliable permeability measurements. In the case of overpressures close to fracture pressures, the permeability is mostly linked to fractures which open by hydrofracturing, and thus the permeability cannot be measured in the laboratory.

The fluid flow out of overpressured compartments to more normally pressured parts of the basin is a function not of the pressure gradients, but of the rate of compaction, which determines the magnitude of the flux (F).

It is therefore difficult to see how rapid upwards flow through sediment with hard overpressure, as suggested by Lynch (1996), can be sustained by fluids from sediment compaction.

The total fluid flux will be a function of the rate of porosity loss and the generation of hydrocarbons from kerogen and dehydration of clay minerals. These processes, which occur at greater burial depth (>2–3 km), are mostly functions of temperature. Without hydrothermal activity it is not possible to heat large volumes of rocks rapidly, because of the high heat capacity of the sediments.

In overpressured compartments, the effective stress is low and the sediments may be overcompacted and then no mechanical compaction occurs. Chemical compaction will, however, continue at a rate that is mainly a function of temperature, but is relatively insensitive to effective stress. In the simple case of constant sedimentation rate and geothermal gradient, the total flux (F) will remain constant. If we also ignore variations in viscosity, it follows from Darcy's law that the pressure gradient is an inverse function of the permeability. Chemical compaction may therefore continue and maintain pressures close to fracture pressures during subsidence at low effective stresses.

Thermal expansion of water has been shown to be relatively insignificant in terms of producing overpressure (Daines 1982), but generation of oil may be important because of both the transformation of solid kerogen to fluid petroleum (Bjørlykke 1993) and the generation of gas (Holm 1996). Dehydration of minerals such as kaolinite and smectite may also be significant. The literature on overpressure has focused on mechanisms increasing the fluid flux, and overpressure has been calculated assuming a certain permeability in the shales. The permeability can, however, at best be predicted within one order of magnitude, and is the major unknown factor in basin modelling (Audet & McConnell 1992). If a shale has a permeability of 10^{-20} instead of 10^{-21} m^2 the compaction-driven flux will be ten times higher while maintaining the same pressure gradient. The permeability is therefore the most important variable determining overpressure. High sedimentation rates will increase the fluid flux and if the permeability-depth function is kept constant, modelling will necessarily indicate that pore pressure is mainly related to sedimentation rates. As sediment compaction is a function of time, high sedimentation rates imply shorter time for compaction, which will cause the porosity and permeability to be higher. In the case of mechanical compaction, high sedimentation rates may cause an overpressure in the mudstones even in otherwise normally pressured basins because it takes time to bleed off the pore water. Chemical compaction is more directly a function of time in addition to temperature: high sedimentation rate will reduce the integrated time–temperature function and cause higher porosity and permeability at a certain depth.

However, the permeability–depth function varies greatly as a function of lithology. Diagenetic reactions are also functions of time, and high sedimentation rates will usually increase the porosity and permeability as a function of depth, thus causing a negative feedback on the pressure build-up.

When fracture pressure is reached, very thin fractures open and the permeability is thus produced by the fluid. The pressure and fluid flux are then independent of the intergranular permeability of the rock matrix.

During this type of chemical compaction the fluid flux is mainly a function of the rate of heating, and this may be taken to be relatively constant for a limited geological time. As discussed above, at a constant fluid flux the pressure gradient is inversely related to the permeability, and when the pressure exceeds the fracture pressure the permeability becomes a dynamic variable, so that small fractures open and adjust their total permeability to the flux required to maintain a certain pressure. As chemical compaction is a function of the rate of heating of the sediments it is not likely that the flow should be episodic.

In the absence of hydrothermal activity the relatively constant heat flux and the large heat capacity of the sediments suggest a rather uniform heating and fluid expulsion.

Studies of modern basins show that development of overpressure is linked to primary facies and to faulting because the lateral drainage is critical (Chapman 1994; Olstad et al. 1997).

During progressive mechanical and chemical compaction sedimentary rocks have ductile properties at low strain rates. This then precludes the formation of open fractures except when fracture pressure is reached (Bjørlykke & Høeg 1997). The ductile properties will also prevent transmission of tectonic anisotropic horizontal stress at low strain rates. During very high overpressures or tectonic uplift sedimentary rocks will have brittle properties and may fracture, but then there is little fluid generation by compaction, as a result of low effective stress and reduced temperature (Bjørlykke & Høeg 1997).

Conclusions

Compaction is still, in most cases, modelled as a function of effective stress, but this can be justified only at shallow depth (<2–3 km depth, <70–100°C). In the case of mechanical compaction, there is a strong negative feedback between the degree of overpressure and the rate of compaction and fluid flow. Pressures close to fracture pressure can therefore not be reached by mechanical compaction because of the low effective stress reducing the rate of compaction and fluid flow.

Compaction-driven fluid fluxes in sedimentary basins are controlled by the rate of porosity reduction and fluid generation in the underlying sequences. In the deeper parts (>2–3 km) of sedimentary basins compaction is mostly controlled by mineralogy and temperature, and may then, for a limited time interval, be relatively constant.

The pressure gradients are inversely related to the permeability, and if the permeability drops below a critical value fracturing will occur. Differences in permeability may focus fluid flow, but do not change the total flux. Both the mechanical and chemical compaction of mudstones depend very much on grain size and mineralogy, and must be linked with primary facies changes and diagenesis.

In subsiding basins mudstones and shales have ductile properties during mechanical and chemical compaction. This limits the potential for generation of fractures except at fracture pressure and for episodic fluid flow.

Modelling of chemical compaction at depth > 2-3 km requires detailed input on the temperature history and the textural and mineralogical composition of the sediments. This may be difficult to model, mostly because of the requirements on the input data. However, this does not justify modelling as a function of effective stress at depths when this is not the main driving force for compaction.

References

APLIN, A. C., YANG, Y. L. & HANSEN, S. 1995. Assessment of beta, the compression coefficient of mudstones and its relationship with detailed lithology. *Marine and Petroleum Geology*, **12**, 955-963.

AUDET, D. M. & McCONNEL, J. D. C. 1992. Forward modelling of porosity and pressure evolution in sedimentary basins. *Basin Research*, **41**, 47-162.

BJØRKUM, P. A. 1995. How important is pressure in causing dissolution of quartz in sandstones. *Journal of Sedimentary Research*, **66**, 147-154.

—— & NADEAU, P. H. 1998. Temperature controlled porosity/permeability reduction, fluid migration and petroleum exploration in sedimentary basins. *Australian Petroleum Production and Exploration Association Journal*, **38**, 453-464.

BJØRLYKKE, K. 1993 Fluid flow in sedimentary basin. *Sedimentary Geology*, **86**, 137-158.

—— 1998. Clay mineral diagenesis in sedimentary basins — a key to the prediction of rock properties. Examples from the North Sea Basin. *Clay Minerals*, **33**, 15-34.

—— & HØEG, K. 1997. Effects of burial diagenesis on stress, compaction and fluid flow in sedimentary basins. *Marine and Petroleum Geology*, **14**, 267-276.

CHAPMAN, R. E. 1994. The geology of abnormal pore pressures. *In*: FERTEL, W. H., CHAPMAN, R. E. & HOTZ, R. F. (eds) *Studies in Abnormal Pressures. Developments in Petroleum Science.* Elsevier, **38**, 19-49.

DAINES, S. R. 1982. Auathermal pressuring and geopressure evaluation. *Bulletin, American Association of Petroleum Geologists*, **66**, 931-939.

HOLM, G. M. 1996. The Central Graben: a dynamic overpressure system. *In*: GLENNIE, K. & HURST, H. (eds) *AD1995: NW Europe's Hydrocarbon Industry.* Geological Society, London, 107-122.

HUNT, J. H. 1990. Generation and migration of petroleum from abnormally pressured fluid compartments. *Bulletin, American Association of Petroleum Geologists*, **74**, 1-2.

ILLIFFE, J. E. & DAWSON, M. R. 1996. Basin modelling history and predictions. *In*: GLENNIE, K. & HURST, H. (eds) *AD1995: NW Europe's Hydrocarbon Industry.* Geological Society, London, 83-105.

LYNCH, F. L. 1996. Mineral/water reaction; fluid flow and Frio Sandstone diagenesis: evidence from the rocks. *Bulletin, American Association of Petroleum Geologists*, **80**, 486-504.

OLSTAD, R., BJØRLYKKE, K. & KARLSEN, D. 1997. Pore water flow and petroleum migration in the Smørbukk field area, offshore Norway. *In*: MØLLER-PEDERSEN, P. & KOESTLER, A. G. (eds) *Hydrocarbon Seals — Importance for Exploration and Production.* Norwegian Petroleum Society (NPF) Publication, 7, 201-216.

PITTMAN, E. D. & LARESE, R. 1991. Compaction of lithic sands: experimental results and applications. *Bulletin, American Association of Petroleum Geologists*, **75**, 1279-1299.

RIEKE, H. H. & CHILINGARIAN, G. V. 1974. *Compaction of Argillaceous Sediments. Developments in Sedimentology 16.* Elsevier, Amsterdam.

WALDERHAUG, O. 1996. Kinetic modelling of quartz cementation and porosity loss in deeply buried sandstone reservoirs. *Bulletin, American Association of Petroleum Geologists*, **80**, 731-745.

YU, Z. H. & LERCHE, I. 1996. Modelling abnormal pressure development in sandstone shale basins. *Marine and Petroleum Geology*, **13**, 179-193.

Permeability anisotropy of consolidated clays

M. B. CLENNELL[1], D. N. DEWHURST[2], K. M. BROWN[3] & G. K. WESTBROOK[4]

[1]*School of Earth Sciences, University of Leeds, Leeds LS2 9JT, UK*
Present address: Centro de Pesquisa em Geofisica e Geologia, IGEO, Universidade Federal da Bahia, Rua Caetano Moura, 123, Salvador, Bahia, 40-190-290, Brazil
(e-mail: clennell@cpgg.ufba.br)
[2]*Department of Geology, Royal School of Mines, Imperial College of Science, Technology and Medicine, Prince Consort Road, London SW7 2BP, UK*
Present address: CSIRO Petroleum, Division of Petroleum Resources, PO Box 3000, Glen Waverley, Victoria, Australia 3150
(e-mail: david.dewhurst@dpr.csiro.au)
[3]*Scripps Institution of Oceanography, Geological Research Division, San Diego, CA, USA*
[4]*School of Earth Sciences, University of Birmingham, Edgbaston, Birmingham B15 2TT, UK*

Abstract: Consolidation of clays tends to result in changes in particle orientation and pore size distribution as well as progressive reduction of porosity and permeability with increasing effective stress. Clay particles are expected to rotate normal to an axial load, thus decreasing flow path tortuosity parallel to the particle alignment direction and increasing tortuosity normal to the particle alignment. This results in the development of anisotropic permeability, such that the horizontal permeability of a consolidated sediment is greater than the vertical permeability at any given porosity. Within any uniform layer, levels of permeability anisotropy are modest. Typically, permeability anisotropy produced by consolidation of natural clays is in the range 1.1–3 and does not reach the high levels predicted by simple models of clay particle reorientation. The discrepancy arises from particle clustering and irregularities in particle packing. Although somewhat higher levels of anisotropy may exist as a consequence of lamination within individual beds, values > 10 that are known to exist on the formation scale are produced by strong contrasts between the permeabilities of interlayered beds. As argillaceous sediments have permeability ranges of many orders of magnitude, apparently subtle lithological layering in a shale unit may lead to a highly anisotropic flow behaviour.

The mechanics of consolidation of argillaceous sediments has been investigated by numerous researchers (e.g. Athy 1930; Griffiths & Joshi 1990; Jones 1994; Aplin et al. 1995; Vasseur et al. 1995). The stress path taken during consolidation determines the alteration of particle, pore and fabric orientations within argillaceous materials, all of which directly affect the movement of fluids in unlithified clay-rich sediments. One-dimensional consolidation with no lateral strain results in the realignment of clay particles as effective stress increases, with a concomitant reduction of the void ratio, e (or porosity, ϕ; $\phi = e/(1 + e)$) of a particular clay (Grunberger et al. 1994; Vasseur et al. 1995). Progressive consolidation leads to the collapse of the largest pores at any given effective stress (Griffiths & Joshi 1990), resulting in grain centres being pushed closer together and therefore increasing the bulk density of the sediment.

Most sedimentary basins contain a high proportion of shales and clay-rich sediments, yet these important constituents are often neglected (assumed not to participate in flow, as in reservoir models) or their permeability is assumed to be homogeneous and assigned a somewhat arbitrary value taken from the literature (e.g. Lambe & Whitman 1979). Natural and remoulded clays show a log–linear decrease in permeability with decreasing void ratio. Data for a wide selection of natural clays and shales collated by Neuzil (1994) suggest that permeability decreases in a log–linear fashion in a band approximately three orders of magnitude wide, over a range of void ratios from 0.1 to 4.0 (c. 10–80% porosity). Permeability of these argillaceous sediments spans eight orders of magnitude, from $10^{-15}\,m^2$ to $10^{-23}\,m^2$. Much of this variation is due to differences in the proportion and particle sizes of the non-clay fraction and in surface properties and aggregated microstructures of different clay minerals (e.g. Olsen 1962; Collins & McGown 1974; Griffiths & Joshi 1989, 1990; Dewhurst et al. 1998).

A further cause of the variation in measured values may be the development of an anisotropy of permeability in clay-rich sediments (Neuzil 1994) as a result of changing particle orientations, pore shapes and pore size distributions. Modelling of fluid flow in clay-rich sediments often neglects the effects of permeability anisotropy, even though values of horizontal permeability exceeding vertical permeability are well documented for unlithified pure and natural clays (e.g. Chan & Kenney 1973; Tavenas et al. 1983a, b; Al-Tabbaa & Wood 1987; Leroueil et al. 1990; Little et al. 1992) and in shales (Young et al. 1964; Neuzil 1994).

Studies of permeability and permeability anisotropy have important implications in several fields, including active margin deformation, brine and hydrocarbon migration in sedimentary basins, geotechnical investigations and hazardous waste disposal. This paper reports a series of geotechnical tests on pure clays and marine clays with a view to documenting the development of permeability anisotropy during consolidation and its possible implications for fluid migration in the subsurface.

Experimental materials

The experimental programme comprised a series of tests on four clays, including kaolinite, a silty clay, calcium montmorillonite and a natural clay recovered on Ocean Drilling Program (ODP) Leg 141 from the Chile Triple Junction accretionary complex (Behrmann et al. 1992). The silty clay used was an artificial test clay, developed at Cornell University to simulate a typical marine clay (Karig & Hou 1992). It comprises 50% silt-grade quartz, 40% illite and 10% smectite with a grain density of 2.68. Liquid limit, LL, and plastic limit, PL, are 58% and 26%, respectively. Commercially available Speswhite kaolin (LL = 69%, PL = 38%) was also tested; it has a grain density of 2.61 and 80% of particles finer than 2 µm diameter. The Ca-montmorillonite used has a grain density of 2.63 and an LL of 120% (determined in 0.05 M $CaCl_2$ solution). The natural clay recovered during ODP Leg 141 (Site 859B, 245 mbsf (m below sea floor)) has a grain density of 2.69, LL of 55% and PL of 21% when remoulded in sea water. X-ray diffraction (XRD) analysis of the clay fraction shows a composition of approximately 20% smectite, 35% illite and 40% chlorite, with minor quartz and feldspar. Bulk-rock XRD analyses indicate dominant quartz, feldspar, illite and chlorite, with lesser amounts of feldspar present (Behrmann et al. 1992).

Two of the tests reported here (kaolinite and silty clay) were conducted using tap water as the permeant. The chemistry of Birmingham water, a mildly alkaline, low-salinity water, is well known from hydrogeological investigations conducted by the University of Birmingham. Distilled or deionized waters are generally unsuitable for the permeability testing of clays, as they may remove weakly bound ions from the clays by leaching, altering the clay chemistry and generating abnormally low permeability as the thickness of the viscous bound water layer becomes much greater as ionic concentration is reduced (Olson & Daniel 1981). Calcium montmorillonite was tested using a 0.05 M solution of $CaCl_2$ as the permeant and cell fluid, to prevent ion exchange reactions and concomitant alteration of clay chemistry, which could affect the clay microstructure and thus the measured values of permeability (Olson & Daniel 1981; Hudec & Yanful 1983). The marine clay from ODP Leg 141 was remoulded and tested using sea water as the permeant to simulate the saline conditions from where it was recovered (e.g. Jose et al. 1989). Initial water contents, W (mass of water/mass of grains), for kaolinite, calcium montmorillonite and the ODP Leg 141 samples were all about 125% of the liquid limits, which is in line with recommendations of Burland (1990) for the water content necessary to disaggregate existing soil structure when remoulding clays. W for the silty clay was 0.6. This discrepancy resulted from the silty clay being part of a separate experiment comparing the permeability evolution of consolidated and sheared clays, in which lower water contents were required (see Dewhurst et al. 1996a, b).

Experimental techniques

The oedometer used in this study has been adapted to measure permeability anisotropy during one-dimensional consolidation (Fig. 1). The basic design allows the permeability to be determined using both constant rate of strain (Smith & Wahls 1969; Wissa et al. 1971; Armour & Drnevich 1986) and constant-rate flow pump techniques (Olsen 1966; Aiban & Znidarcic 1989; Little et al. 1992). Flow rates are considered to be accurate to 1% at the mid-range, and 5% at the lowest range. The flow pump technique has several advan-

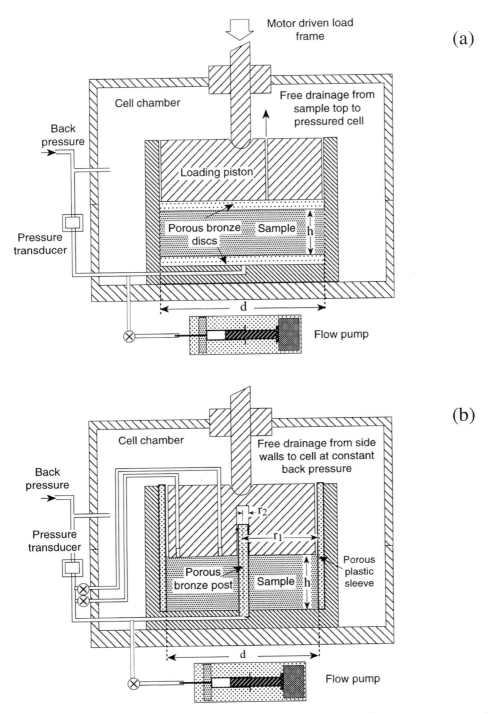

Fig. 1. (a) Oedometric permeameter for tests conducted with vertical drainage. (b) Oedometric permeameter for tests conducted with radial drainage.

tages over the standard constant head and falling head methods of permeability measurement, the most important being the ease with which very low flow rates (and hence low hydraulic gradients) can be reliably maintained (Olsen 1966; Pane et al. 1983; Olsen et al. 1985, 1988). Other advantages include accuracy of the technique, shorter testing times and the fact that a small volume of pore fluid is used; this ensures that sample disturbance during testing is minimal. Measurements of permeability using the constant rate of strain and flow pump tests gave almost identical results (Dewhurst et al. 1996a). In the radial drainage configuration (Fig. 1b), a porous bronze post is inserted through the centre of the sample chamber and the perimeter is lined with a stiff porous plastic membrane of 3 mm thickness. Radial permeability measurements are made by pumping the permeant from the central porous post to the sample edge.

The oedometer is contained within a pressurized triaxial cell and has two configurations that allow either vertical or radial drainage during load increments and permeability tests. Drainage is against a constant back-pressure of 300 kPa, generated by a mercury pot system, necessary to keep the sample fully saturated (Lowe et al. 1964). During consolidation tests using the vertical drainage configuration (Fig. 1a), a thin screen of filter paper is used between the sample and the porous bronze discs to stop the latter clogging with clay particles.

The remoulded sample is placed in the chamber and de-aired before loading by immersion of the entire sample cylinder and piston under vacuum. The oedometer is then immersed in de-aired water within the triaxial cell and the back-pressure applied. The experiment is left for 24 h to allow trapped air to be driven into solution. The clays are loaded at a constant rate of displacement (10^{-2} mm min^{-1}) with a progressive load increment ratio of unity (i.e. effective stress was doubled for each increment). That is, each sample is subjected to a series of increasing vertical loads, each increment being applied gradually by using the load frame to move the piston at a low constant rate until the required total stress is reached. At the end of each increment, the system is allowed to equilibrate for 24 h. In this time, any excess pore pressure dissipates and the piston creeps somewhat, after which the vertical effective stress is noted (total vertical stress minus back-pressure). At this point, a constant-rate-of-flow permeability test was performed by infusing the permeant from the syringe pump. Flow-induced excess pressure at the inflow point is monitored relative to the constant back-pressure by a differential pressure transducer. Hydraulic equilibrium was reached in a period ranging from 30 min to 24 h, during which time a volume of between 0.2 and 1.0 ml of permeant had been infused for each test.

After testing, the final water content of the clay was measured by weighing a sub-sample wet, and after oven drying at 105°C for 24 h. Final porosity was calculated from this water content and the known grain density, and back-calculated for the other increments in the test series using the measurements of sample thickness given by a dial gauge recording piston displacement to 0.01 mm.

Results

Consolidation tests

Figures 2–5 show the results of the consolidation tests on these clays. Kaolinite showed a characteristic log–linear decrease in equilibrium void ratio (porosity, more familiar to geologists, is plotted on the right-hand scale in each figure) with increasing vertical effective stress in both the vertical and radial drainage configurations (Fig. 2a). The volumetric curves from the two configurations conform closely to each other, suggesting that the decrease in equilibrium void ratio is independent of the drainage path through both the loading and unloading cycles (some of the data on the unloading path at effective stresses < 1800 kPa for radially drained kaolinite were lost as a result of a fault in the data recording system). The plot of the vertical and radial permeabilities against void ratio is illustrated in Fig. 2b and shows that log–linear relationships were established during consolidation. Although at high void ratios there is very little anisotropy between the two flow directions, these relationships have slightly different gradients. This leads to the steady divergence of the curves, ultimately leading to a permeability anisotropy of c. 1.7 developing at a void ratio of 0.7 (vertical effective stress at this point was c. 3 MPa). These results are similar in magnitude and trend to those obtained by Al-Tabbaa & Wood (1987) and Aiban & Znidarcic (1988).

The effective stress–void ratio curves for the silty clay (Fig. 3a) show the typical log–linear decrease in void ratio with increasing average vertical effective stress. In contrast to the curves for kaolinite, these curves do not coincide with one another, even though the samples were prepared with equal initial water contents. We believe that this is due to slight differences in pore-water loss during the emplacement of the top platen on the sample, to ensure that an air pocket does not form between the sample and the platen. The decrease of permeability with decreasing void ratio (Fig. 3b) is again log–linear (see Neuzil 1994) for both the vertical and radial configurations, and illustrates the lack of permeability anisotropy developed in the silty clay, even at the low void ratios (0.35–0.40) attained at the maximum stress levels (c. 4 MPa). During radial permeability testing, axial loading of the sample may lead to the development of zones of vertical particle alignment at the radial drainage boundaries; this process is termed smearing and may reduce the radial permeability by increasing the tortuosity of flow paths near the drainage boundaries (Wilkinson & Shipley 1972). After

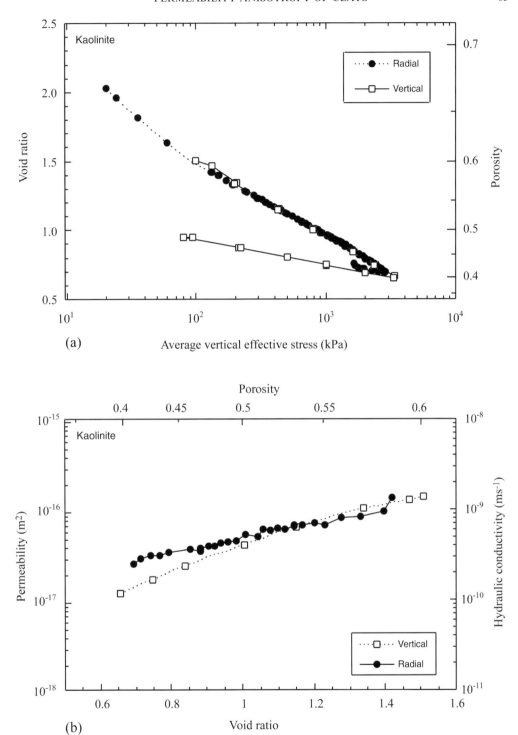

Fig. 2. (a) Effective stress–void ratio plot for consolidated kaolinite. (b) Void ratio–permeability plot for consolidated kaolinite, showing an increase in permeability anisotropy from one to 1.7 as void ratio decreases. The symbols used are consistent with those in Figs 3–5.

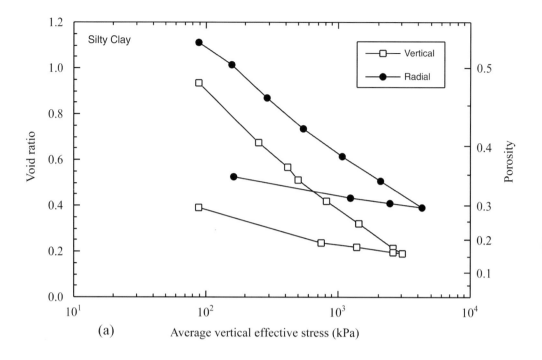

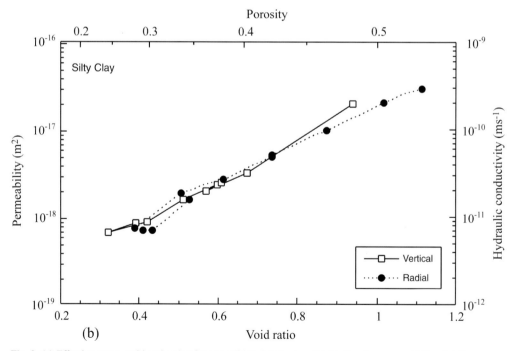

Fig. 3. (a) Effective stress–void ratio plot for consolidated silty clay. (b) Void ratio–permeability plot for consolidated silty clay, illustrating the lack of permeability anisotropy in this clay.

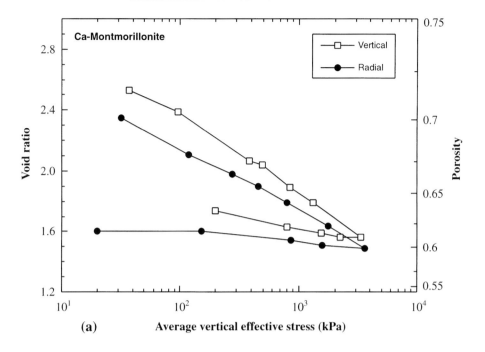

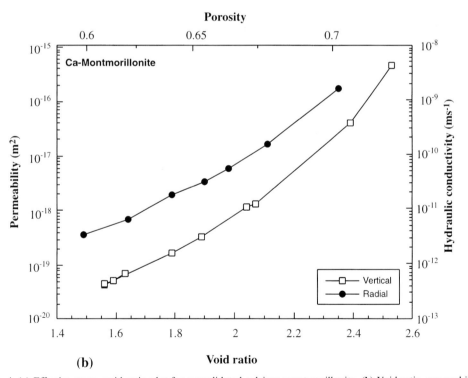

Fig. 4. (a) Effective stress–void ratio plot for consolidated calcium montmorillonite. (b) Void ratio–permeability plot for consolidated calcium montmorillonite, reaching a maximum permeability anisotropy of eight. Comparison with the other figures shows this clay has the lowest permeability measured among these sediments and also the highest void ratio, indicative of the fine grain size and thus small pores in this sediment.

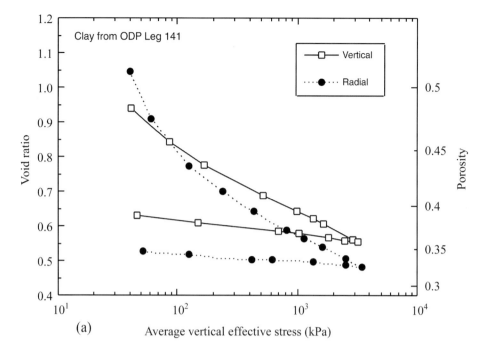

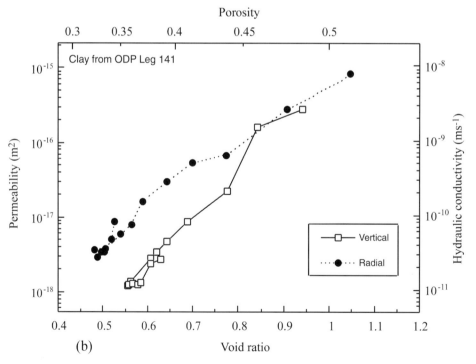

Fig. 5. (a) Effective stress–void ratio plot for a consolidated clay from ODP Leg 141. (b) Void ratio–permeability plot for a consolidated clay from ODP Leg 141, showing a maximum permeability anisotropy of approximately five.

the end of the test cycle, the microstructures of the clay at the side wall and post were examined optically and with scanning electron microscopy (SEM) techniques. Smearing, although present, was confined to an extremely thin layer (<10 μm) a few clay plates thick, penetrated by large pores. These observations indicate that smearing did not adversely affect permeability measurements of these samples.

The log–linear decrease in equilibrium void ratio with increasing vertical effective stress for vertically and radially drained calcium montmorillonite is shown in Fig. 4a. Similar to those for the silty clay, the curves for loading and unloading are not coincident, with radially drained montmorillonite exhibiting a higher void ratio at any given effective stress. Anisotropy of permeability developed in consolidated calcium montmorillonite develops with the onset of consolidation and increases from four, at a void ratio of $c.$ 2.4, to eight at a void ratio of $c.$ 1.6 (Fig. 4b). The vertically drained sample of montmorillonite exhibits the lowest permeability (4×10^{-20} m^2) of the four clays tested at the peak effective stress (3.5 MPa) even though it has the highest void ratio (1.6) at this point.

The load–void ratio consolidation curves for the natural silty clay from the Chile Triple Junction are shown in Fig. 5a. Similar to the curves for montmorillonite and the artificial silty clay, the load–void ratio curves for the vertical and radial drainage configurations are not coincident for this clay. The development of permeability anisotropy for this clay can be observed in Fig. 5b; the permeability ratio is $c.$ 1 at void ratios >0.8, but increases to $c.$ 5 at void ratios <0.6.

Microstructural observations

Each sample was recovered after the end of a complete test series, so the fabrics studied were developed following the attainment of peak stresses and subsequent elastic unloading. SEM in secondary electron mode was used on broken surfaces prepared from the test specimens. To avoid shrinkage, cracking and distortion, samples were either air-dried slowly or critical-point dried under a CO_2 atmosphere, the preparation method varying according to the mineralogy. Standard thin sections of resin-impregnated test samples were also examined in normal and cross-polarized light, and using a sensitive tint plate to pick out preferred orientations. Polished blocks of impregnated material

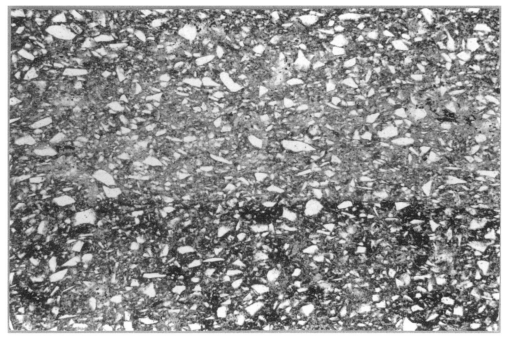

Fig. 6. Photomicrograph of a thin section of consolidated silty clay illustrating some alignment of the long axes of irregularly shaped silt particles normal to the direction of applied stress (width of field of view is $c.$ 3 mm). All the micrographs illustrated are vertical sections through the clays.

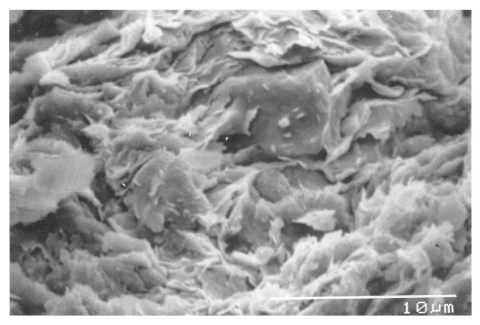

Fig. 7. Photomicrograph illustrating clay particles wrapping around silt grains, which may result in enhanced (i.e. differential) compaction. Scale bar represents 10 μm.

were also examined in the electron microscope in back-scattered mode.

Consolidated samples of artificial silty clay at low magnifications show a clear sub-horizontal alignment of larger, irregularly shaped silt particles, imaged in optical micrographs (Fig. 6). Grain breakage is occasionally noted in the quartz particles. The optical properties under a polarizing microscope show that the bulk of the clay had a weak mean foliation aligned horizontally. At high magnifications using SEM, clay particles exhibit differential compaction around silt particles (Fig. 7). Some small clay domains were moderately aligned; however, although some of these domains were oriented normal to the axial stress, many were aligned at moderate angles (10–30°) to this orientation. The whole fabric becomes more compact at higher loads, the larger pores between domains of clay and silt fragments being reduced in number and size at the highest loads. The overall degree of grain alignment does not appear to increase significantly at loads over 250 kPa. This is in agreement with other studies on natural and artificial clays (e.g. McConnachie 1974; Bennett et al. 1989). Some edge effects are noted, but these are mainly artefacts associated with the removal of cohesive, plastic clay from the oedometer.

Kaolinite, comprising well-sorted robust and platy grains with a moderate aspect ratio of about 15:1, produces strong consolidation fabrics during oedometric loading, which are observable at magnifications of 100× to 5000×. The degree of grain alignment in the photomicrographs of thin-sectioned kaolinite, although not everywhere intense, is generally strong (Fig. 8). Computer analysis of representative fields of view c. 200 μm across give a consistency ratio (a statistical measure of orientation; see Smart & Leng 1993) of about 95% for a preferred orientation within 5° of horizontal (Clennell et al., unpublished results). The microstructure is hierarchical rather than uniform, consisting of roughly polygonal domains a few hundreds of microns across separated by thin inter-domain areas (Figs 9 and 10). Particles within clay domains are strongly aligned, but the orientation within each diverges somewhat from its neighbours, and the overall alignment may be slightly bowed or kinked. The inter-domain areas are either narrow zones of less ordered and more open packing, or well-aligned regions with a more compact fabric oriented oblique to the alignment of the larger domains (Fig. 10). These structures appear to be shear zones accommodating slip between the domains, which may compact at slightly different rates.

Edge effects were significant in the sample: at the side walls in both tests and next to the central drainage post some smearing of kaolin grains occurred, leading to a zone of sub-vertical fabric some 100–300 μm thick. This smearing, which is

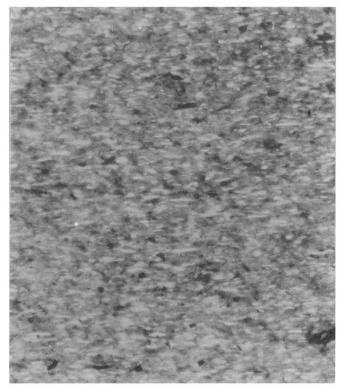

Fig. 8. Thin section of consolidated kaolinite under crossed polars illustrating a strong preferential orientation of clay particles normal to the direction of applied stress at low magnifications. Width of field of view is c. 0.2 mm.

controlled mainly by the larger and more rigid kaolin particles, may account for the unexpected similarity of vertical and radial permeability at high void ratios. Edge effects are not evident at the top and base of the sample, nor is any marked change noted in the overall microstructure, progressing through the sample from top to bottom or centre to edge (apart from the smeared zone) in either series of tests. It is therefore probable that sample disturbance caused by uneven loading or pressure distribution, or by seepage-induced consolidation (Pane et al. 1983) was minimal.

Consolidated, remoulded natural clay from the Chile Triple Junction comprises poorly sorted clay and silt-sized grains and generally exhibits no clear alignment of particles at any scale of observation. Occasional discrete zones of particle alignment are noted, with orientations 10–20° from horizontal. The fabric, in general, retains an open structure at both low (100 kPa) and high (3.2 MPa) loads (Figs 11 and 12, respectively). In contrast to the artificial silty clay, enhanced consolidation of clay particles around silt grains is not observed in this sample; this is particularly noticeable in pits resulting from the plucking out of silt grains during sample preparation, where no significant change in particle orientation or spacing is observed. Smearing was detected at the inner and outer radial drainage boundaries but, similarly to the artificial silty clay, the zone is at most 15 μm thick, and is penetrated by large pores and so is unlikely to affect the permeability results.

Imaging the calcium montmorillonite used in this experimental programme proved extremely difficult. Intense shrinkage on drying resulted in the formation of ubiquitous, intersecting desiccation cracks and large topographic features on broken surfaces. Critical-point dried material was friable and the samples tended to disintegrate in the vacuum of the scanning electron microscope observation chamber. Furthermore, these samples suffered from intense charging, which resulted in extreme problems for obtaining meaningful photomicrographs using either secondary or back-scattered electron imaging. A strong preferential orientation of horizontally aligned individual particles (0.1–0.4 μm) and aggregates of particles (2–20 μm) could be

Fig. 9. Secondary electron images of consolidated kaolinite illustrating that the fabric does not simply consist of aligned individual clay plates. Aggregations of grains are clearly visible in both micrographs; a slip surface is noted down the right-hand side of the upper micrograph (even though the consolidation tests allowed no lateral strain). The large particles in the centre of the lower micrograph are lying on the broken surface of the sample. Scale bars represent 20 μm in both micrographs.

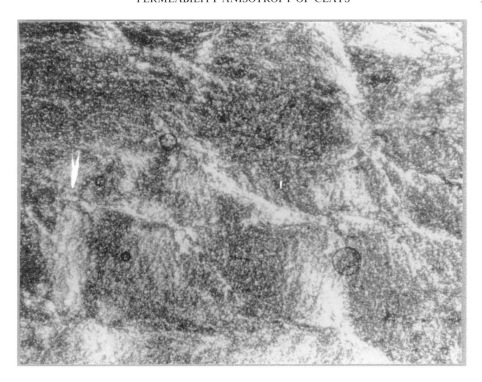

Fig. 10. Thin sections of consolidated kaolinite under crossed polars, showing considerable heterogeneity of the fabric. Slip surfaces separating domains of kaolinite (white bands) are visible in the upper micrograph (width of field of view is c. 1.5 mm), whereas aligned domains of clay plates with orientations divergent with respect to one another are visible in the lower micrograph (width of field of view is c. 0.4 mm). Figure 8 is a close-up from within one of these domains happening to have nearly perfect horizontal alignment.

Fig. 11. Secondary electron image of a clay from ODP Leg 141, loaded to c. 100 kPa. The clay has an open fabric and a high porosity, with little fabric alignment evident. Scale bars represent 10 μm.

picked out, but further attempts to observe and analyse the images obtained from this clay were unsuccessful.

Discussion

Microstructure and its control on the magnitude of permeability anisotropy

The development of anisotropic permeability in consolidated clays is generally ascribed to grain reorientation and preferential alignment of particles. Remoulded clays and naturally occurring marine clays have high initial porosities associated with small, randomly oriented clay domains that enclose large irregular voids (Bennett & Hulbert 1986). As burial and consolidation proceed, these open fabrics give way to increasingly better aligned clay domains with thin, long, narrow voids. It is expected intuitively that an increasing vertical effective stress (under conditions of zero lateral strain) would produce an increasing horizontal alignment of platy clay grains and so a more rapid decrease in vertical permeability than in horizontal permeability.

The purpose of the microfabric investigations was to relate the structures visible at all scales to the measured absolute values of permeability (largely controlled by pore-throat sizes and connectivity) and to the degree of permeability anisotropy (controlled largely by orientation of fabric and continuity of structural elements). Specifically, it was hoped to relate the degree of grain alignment produced by vertical loading in the oedometer to the changes in permeability of the clay. Conceptual models (e.g. Olsen 1962) would suggest that consistent, preferential orientation of particles with aspect ratios similar to those of kaolinite and montmorillonite would lead to high levels of permeability anisotropy. Following from this, Arch & Maltman (1990) used two-dimensional, computer-generated arrangements of platy grains to predict that in sediment fabrics with a high degree of clay grain parallelism, permeability anisotropies of kaolinite and montmorillonite should reach values of around 20 and >100, respectively. In such models the flow paths parallel to the grain alignment are relatively straight, whereas those running across the fabric are much more sinuous. This contrast in tortuosity increases as the width to thickness ratio of the clay plates increases. Permeability is assumed to scale with the square of tortuosity in each direction, so that increasing fabric orientation magnifies the permeability anisotropy. Accordingly, these effects should be greatest in materials with a high proportion of platy clay minerals.

Fig. 12. Secondary electron image of a clay from ODP Leg 141, loaded to $c.$ 3200 kPa. Comparison with Fig. 11 shows that the clay still retains an open structure, although the porosity has reduced from 50% at 100 kPa to 33% at 3200 kPa. Little fabric alignment is evident, and what there is is not horizontal, even though this clay has been subjected to uniaxial vertical loading and a much higher effective stress than that in Fig. 11. Scale bars represent 10 µm.

The low levels of permeability anisotropy we found in the artificial silty clay are consistent with a large body of data on the permeability of natural silty clays (e.g. Garcia-Bengoechea et al. 1979; Garcia-Bengoechea & Lovell 1981; Shephard & Bryant 1983; Tavenas et al. 1983a, b; Lapierre et al. 1990; Leroueil et al. 1990; Taylor & Leonard 1990). We would not expect these mixtures to have highly anisotropic permeability because there is little potential for strong grain alignment.

In the pure clays, high levels of fabric anisotropy were recorded, and in the case of kaolin this can be quantified by image analysis. However, the overall level of anisotropy in permeability remained modest even to high degrees of consolidation. For kaolin, the level of anisotropy reached in this and other studies (e.g. Al-Tabbaa & Wood 1987; Znidarcic & Aiban 1988) is generally 2–3 whereas for montmorillonite, it did not exceed eight. Furthermore, the sample from the Chile Triple Junction, which had only a weak foliation visible in micrographs, had, relatively, highly anisotropic permeability.

Clearly, the directional permeability is controlled by factors other than the average level of particle orientation. Olsen (1962) suggested that the discrepancy was due to particle clustering, so that although intra-aggregate pores have strongly anisotropic tortuosity, the dominant flow channels though the inter-aggregate zones have weaker preferred orientation. The deformable cluster theory was further developed by Yong & Warkentin (1975) and Mitchell (1993). Evidence for bimodal pore-size distributions in which the larger, inter-aggregate pores collapse during consolidation comes from mercury injection measurements (Garcia-Bengoechea et al. 1979; Garcia-Bengoechea & Lovell 1981; Delage & Lefebvre 1984; Griffiths & Joshi 1990).

However, although clustering and aggregation of particles does occur at the small scale, particularly in marine sediments where flocculation occurs, it was not observed in photomicrographs of the silty clay or kaolinite, which were remoulded in tap water. The degree of grain alignment observed in electron micrographs and optical sections of kaolinite is strong; however, the packing, rather than being homogeneous, has an irregular structure with domains separated by small shear zones. Particle alignment in shear orientations in consolidated kaolinite has also been documented by Grunberger et al. (1994) and Vasseur et al. (1995). The oblique fabrics illustrated in Figs 9 and 10 are probably local shear zones accommodating heterogeneous consolidation within the sample. Together with small edge effects, these heterogeneities may combine to reduce the overall level of anisotropy in the permeability from that predicted by the uniform packing models of Olsen (1962) and Arch & Maltman (1990).

We would suggest that there are more fundamental reasons why sediments do not acquire high levels of permeability anisotropy. In consolidation experiments on pure clays, an increase in clay particle orientation can be demonstrated with increasing load, but most orientation occurs in the very early stages of loading, while the material is still a slurry. In the case of natural clay sediments, increased alignment as a result of mechanical orientation can be demonstrated for the first few metres of burial (Bennett & Hulbert 1986). Many natural sediments have components of strength and stiffness, termed 'structure', that cannot be attributed to the reduction in porosity during the consolidation history of the sediment. These properties arise from deposition of matter from solution (e.g. carbonates, hydroxides and organic material), mineral recrystallization, inter-particle attractive forces and modification of the adsorbed water layer (Burland 1990; Leroueil & Vaughan 1990). This structure must collapse before particles can behave independently and acquire a high degree of orientation: studies of particle orientation of natural sediments as a function of burial depth show little increase in orientation beyond a few tens of metres of burial. Beyond this depth an increase in clay particle alignment is discernible in some natural clays, but it is often obscured by the scatter of data (Bennett et al. 1991, and references therein).

Although we accept that clay plates can align strongly, there remain some serious conceptual problems with the simple model of grain alignment and its relationship to permeability ratio. According to Witt & Brauns (1983), the tortuosity ratio of randomly packed ellipsoidal grains cannot exceed 2.5, even if the grains are allowed to approach the shape of flat plates by increasing their aspect ratios to those typical of clays (4–100). Following the arguments of Witt & Brauns, even with certain (unlikely) non-random arrangements that produce exceptional levels of tortuosity, it is not feasible to obtain very high levels of anisotropy in a homogeneous material. Essentially, the material must become inhomogeneous, i.e. layered, to produce a high degree of permeability anisotropy. To obtain a permeability ratio of ten in a layered soil, it is necessary to have a highly segregated laminar structure with the intercalated laminae having markedly different permeabilities (Chan & Kenney 1973). Only a few natural clays and varved silty sediments have anisotropy ratios approaching this value. Most sediments, even those deeply buried, have permeability ratios ranging from one to 2.5: values as high as four are exceptional (Freeze & Cherry 1978; Al-Tabbaa & Wood 1987; Chandler et al. 1990; Leroueil et al. 1990).

Mean pore size and pore-throat size decrease during consolidation of clays (e.g. Griffiths & Joshi 1990). Therefore, the permeability of a clay will decrease rapidly, even if the total cross-section of pore openings available for flow falls only slightly. The bulk of the flow through clays tends to be partitioned preferentially into a few, larger diameter flow channels (Delage & Lefebvre 1984), with the numerous smaller pores (which may together constitute a greater proportion of the total pore space) contributing much less to the transmission of fluid. Considering the electron micrographs of consolidated kaolin, for example, the largest pores have varying shapes and are the least oriented, whereas the smallest pores are the most oriented and most anisotropic in shape. It thus seems likely that pore orientation and the size of the largest pores will control the permeability and the development of permeability anisotropy in consolidating clays.

Conclusions

The results of these tests extend those of previous studies at low effective stress (<1 MPa), indicating that uniaxial consolidation alone produces very little permeability anisotropy. Tests on pure clays attained anisotropies of 1.7 in kaolinite and eight in montmorillonite. Levels of absolute permeability and its anisotropy following consolidation of an artificial silty clay were comparable with those reported for similar natural materials. Uniaxial consolidation alone could not produce any measurable anisotropy of permeability in this clay despite the fact that a consistently anisotropic grain fabric developed during progressive loading. Conversely, a natural, marine silty clay attained a permeability anisotropy of five without any clear fabric anisotropy being developed. These findings, though somewhat counter-intuitive, are consistent with a large body of laboratory data suggesting that consolidation-induced anisotropic permeability is a limited phenomenon in silt-bearing clays.

Several factors reduce permeability anisotropy in clays to values lower than those predicted by simple models of particle orientation. The plate arrangements in the idealized models may be improbable or impossible given the constraints that real clay plates have a range of sizes and aspect ratios, and must form a load-bearing skeleton. Second, real flow paths are not as angular or sinuous as those in the ideal models, and flow is concentrated through the larger pores. Furthermore, the shape of the pores does not mimic the shape of the grains, so high levels of grain alignment do not necessarily lead to high levels of pore alignment.

There is little likelihood of stronger alignment, and so greater anisotropy, in homogeneous clay soils and sediments encountered in the field. Anisotropy of permeability may, however, be achieved on a much larger scale by the interbedding of strata of highly variable permeability, such as varved sediments, or larger-scale alternations of sandstone and shale.

We are very grateful to the technical and academic staff of the School of Civil Engineering at the University of Birmingham, in whose research laboratory this work was carried out. The technical expertise of J. Clarke is also acknowledged for the manufacture of the permeameters. The artificial silty clay was kindly donated by D. Karig of Cornell University, the Speswhite kaolinite by the ECC Ltd and the montmorillonite by Laporte Absorbents Ltd. The work was funded by NERC Ocean Drilling Program Special Topic Grant GST/02/551.

References

AIBAN, S. A. & ZNIDARCIC, D. 1989. Evaluation of the flow-pump and constant head techniques for permeability measurements. *Géotechnique*, **39**, 655–666.

AL-TABBAA, A. & WOOD, D. M. 1987. Some measurements of the permeability of kaolin. *Géotechnique*, **37**, 499–503.

APLIN, A. C., YANG, Y. & HANSEN, S. 1995. Assessment of β, the compression coefficient of mudstones and its relationship with detailed lithology. *Marine and Petroleum Geology*, **12**, 955–963.

ARCH, J. & MALTMAN, A. J. 1990. Anisotropic permeability and tortuosity in deformed wet sediments. *Journal of Geophysical Research*, **95**, 9035–9047.

ARMOUR, D. W. & DRNEVICH, V. P. 1986. Improved techniques for the constant-rate-of-strain consolidation test. *In*: YONG, R. N. & TOWNSEND, F. C. (eds) *Consolidation of Soils, Testing and Evaluation*. American Society for Testing and Materials, STP **892**, 170–183.

ATHY, L. F. 1930. Density, porosity and consolidation of sedimentary rock. *Bulletin, American Association of Petroleum Geologists*, **31**, 241–287.

BEHRMANN, J. H., LEWIS, S. D., MUSGRAVE, R. J. et al. 1992. *Proceedings of the Ocean Drilling Program, Initial Reports, 141*. Ocean Drilling Program, College Station, TX.

BENNETT, R. H. & HULBERT, M. H. 1986. *Clay Microstructure*. Geological Science Series, International Human Resources Development Corporation, Boston, MA.

——, BRYANT, W. R. & HULBERT, M. H. 1991. *Microstructures of Fine Grained Sediments from Mud to Shale*. Springer, New York.

——, FISCHER, K. M., LAVOIE, D. L., BRYANT, W. R. & REZAK, R. 1989. Porometry and fabric of marine clay and carbonate sediments: determinants of permeability. *Marine Geology*, **89**, 127–152.

BURLAND, J. B. 1990. On the compressibility and shear strength of natural clays. *Géotechnique*, **40**, 329–378.

CHAN, H. T. & KENNEY, T. C. 1973. Laboratory investigation of the permeability ratio of New Liskeard varved soil. *Canadian Geotechnical Journal*, **10**, 453–472.

CHANDLER, R. J., LEROUEIL, S. & TRENTER, N. A. 1990. Measurements of the permeability of London Clay using a self-boring permeameter. *Géotechnique*, **40**, 113–124.

COLLINS, K. & MCGOWN, A. 1974. The form and function of microfabric features in a variety of natural soils. *Géotechnique*, **24**, 223–254.

DELAGE, P. & LEFEBVRE, G. 1984. Study of the structure of a sensitive Champlain clay and its evolution during consolidation. *Canadian Geotechnical Journal*, **21**, 21–35.

DEWHURST, D. N., APLIN, A. C., SARDA, J. P. & YANG, Y. 1998. Compaction-driven evolution of porosity and permeability in natural mudstones: an experimental study. *Journal of Geophysical Research*, **103**, 651–661.

——, BROWN, K. M., CLENNELL, M. B. & WESTBROOK, G. K. 1996a. A comparison of the fabric and permeability anisotropy of consolidated and sheared silty clay. *Engineering Geology*, **42**, 253–267.

——, CLENNELL, M. B., BROWN, K. M. & WESTBROOK, G. K. 1996b. Fabric and hydraulic conductivity of sheared clays. *Géotechnique*, **46**, 761–768.

FREEZE, R. A. & CHERRY, J. A. 1978. *Groundwater*. Prentice–Hall, Englewood Cliffs, NJ.

GARCIA-BENGOECHEA, I. & LOVELL, C. W. 1981. Correlative measurements of pore size distribution and permeability in soils. *In*: ZIMMIE, T. F. & Riggs, C. O. (eds) *Permeability and Groundwater Contaminant Transport*. American Society for Testing and Materials, STP, **746**, 137–150.

——, —— & ALTSCHAEFFL, A. G. 1979. Pore size distribution and permeability of silty clays. *Journal of the Geotechnical Engineering Division, American Society of Civil Engineers*, **105**, 839–856.

GRIFFITHS, F. J. & JOSHI, R. C. 1989. Change in pore size distribution due to consolidation of clays. *Géotechnique*, **39**, 159–167.

—— & —— 1990. Clay fabric response to consolidation. *Applied Clay Science*, **5**, 37–66.

GRUNBERGER, D., DJÉRAN-MAIGRE, I., VELDE, B. & TESSIER, D. 1994. Measurement through direct observation for kaolinite particle reorientation during compaction. *Comptes Rendus de l'Académie des Sciences*, **318**, 627–633.

HUDEC, P. P. & YANFUL, E. K. 1983. Properties of brine-treated clays. *In*: SEKI, Y. (ed.) *Proceedings, Fourth International Conference on Water–Rock Interaction, Miasa, Japan*, 191–194.

JONES, M. 1994. Mechanical principles of sediment deformation. *In*: MALTMAN, A. J. (ed.) *Geological Deformation of Sediments*. Chapman and Hall, London.

JOSE, U. V., BHAT, S. T. & NAYAKA, B. U. 1989. Influence of salinity on permeability characteristics of marine sediments. *Marine Geotechnology*, **8**, 249–258.

KARIG, D. E. & HOU, G. 1992. High-stress consolidation experiments and their geologic implications. *Journal of Geophysical Research*, **97**, 289–300.

LAMBE, T. W. & WHITMAN, R. V. 1979. *Soil Mechanics, SI version*. Wiley, New York.

LAPIERRE, C., LEROUEIL, S. & LOCAT, J. 1990. Mercury intrusion and permeability of Louiseville clay. *Canadian Geotechnical Journal*, **27**, 761–773.

LEROEUIL, S. & VAUGHAN, P. 1990. The general and congruent effects of structure in natural soils and weak rocks. *Géotechnique*, **40**, 467–488.

——, BOUCLIN, G., TAVENAS, F., BERGERON, I. & LA ROCHELLE, P. 1990. Permeability anisotropy of natural clays as a function of strain. *Canadian Geotechnical Journal*, **27**, 568–579.

LITTLE, J. A., MUIR-WOOD, D., PAUL, M. A. & BOUAZZA, A. 1992. Some laboratory measurements of Bothkennar clay in relation to soil fabric. *Géotechnique*, **42**, 355–361.

LOWE, J., ZACCHEO, P. F. & FELDMAN, H. S. 1964. Consolidation testing with back pressure. *Journal of the Geotechnical Engineering Division, American Society of Civil Engineers*, **90**, 69–86.

MCCONNACHIE, I. 1974. Fabric changes in consolidated kaolin. *Géotechnique*, **24**, 207–222.

MITCHELL, J. K. 1993. *Fundamentals of Soil Behaviour*, 2nd edn. Wiley, New York.

NEUZIL, C. E. 1994. How permeable are clays and shales? *Water Resources Research*, **30**, 145–150.

OLSEN, H. W. 1962. Hydraulic flow through saturated clays. *Clays and Clay Minerals*, **9**, 131–162.

—— 1966. Simultaneous fluxes of liquid and charge in saturated kaolinite. *Proceedings of the American Soil Science Society*, **33**, 338–344.

—— & DANIEL, D. E. 1981. Measurement of the hydraulic conductivity of fine grained soils. *In*: ZIMMIE, T. F. & RIGGS, C. O. (eds) *Permeability and Groundwater Contaminant Transport*. American Society for Testing and Materials, STP, **746**, 18–64.

——, MORIN, R. H. & NICHOLS, R. W. 1988. Flow pump applications in triaxial testing. *In*: DONAGHE, R. T., CHANEY, R. C. & SILVER, M. L. (eds) *Advanced Triaxial Testing of Soil and Rock*. American Society for Testing and Materials, STP, **977**, 68–81.

——, NICHOLS, R. W. & RICE, T. L. 1985. Low gradient permeability measurements in a triaxial system. *Géotechnique*, **35**, 145–157.

PANE, V., CROCE, P., ZNIDARCIC, D., HO, H. Y., OLSEN, H. W. & SCHIFFMAN, R. L. 1983. Effects of consolidation on permeability measurements for soft clay. *Géotechnique*, **33**, 67–72.

SHEPHERD, L. E. & BRYANT, W. R. 1983. Geotechnical properties of lower trench, inner slope and outer slope sediments. *Tectonophysics*, **99**, 279–312.

SMART, P. & LENG, X. 1993. Present developments in image analysis. *Scanning Microscopy*, **7**, 5–16.

SMITH, R. E. & WAHLS, H. E. 1969. Consolidation under constant rates of strain. *Journal of the Soil Mechanics and Foundations Division, ASCE*, **95**, 519–539.

TAVENAS, F., JEAN, P., LEBLOND, P. & LEROUEIL, S. 1983a. The permeability of natural soft clays. Part 2: Permeability characteristics. *Canadian Geotechnical Journal* **20**, 645–660.

——, LEBLOND, P., JEAN, P. & LEROUEIL, S. 1983b. The permeability of natural soft clays. Part 1: Methods of laboratory measurement. *Canadian Geotechnical Journal*, **20**, 629–644.

TAYLOR, E. & LEONARD, J. 1990. Sediment consolidation and permeability at the Barbados forearc. *In*: MOORE, J. C., MASCLE, A. *et al.* (eds) *Proceedings of the Ocean Drilling Program, Scientific Results, 110*. Ocean Drilling Program, College Station, TX, 289–308.

VASSEUR, G., DJERAN-MAIGRE, I., GRUNBERGER, D., TESSIER, D., ROUSSET, G. & VELDE, B. 1995. Evolution of structural and physical parameters of clays during experimental compaction. *Marine and Petroleum Geology*, **12**, 945–955.

WILKINSON, W. B. & SHIPLEY, E. L. 1972. Vertical and horizontal laboratory permeability measurements in clay soils. *Developments in Soil Science 2. Fundamentals of Transport through Porous Media*. I.A.H.K. Elsevier, Amsterdam, 285–298.

WISSA, A. E. Z., CHRISTIAN, J. T., DAVIS, E. H. & HEIBURG, A. M. 1971. Consolidation at constant rate of strain. *Journal of the Soil Mechanics and Foundations Division, ASCE*, **97**, 1393–1412.

WITT, K. J. & BRAUNS, J. 1983. Permeability anisotropy due to particle shape. *Journal of Geotechnical Engineering*, **109**, 1181–1187.

YONG, R. N. & WARKENTIN, B. P. 1975. *Soil Properties and Soil Behaviour. Developments in Geotechnical Engineering, 5*. Elsevier, Amsterdam.

YOUNG, A., LOW, P. F. & MCLATCHIE, A. S. 1964. Permeability studies of argillaceous rocks. *Journal of Geophysical Research*, **69**, 4237–4245.

ZNIDARCIC, D. & AIBAN, S. A. 1988. Comment on Al-Tabbaa and Wood: 'Some measurements of the permeability of kaolin'. *Géotechnique*, **38**, 453–454.

Insights into the hydraulic performance of landfill-lining clays during deformation

M. G. PETERS & A. J. MALTMAN

Institute of Geography and Earth Sciences, University of Wales, Aberystwyth SY23 2DB, UK

Abstract: The low hydraulic conductivities of mudrocks are widely utilized in the waste disposal industry, where clays are commonly employed as a lining to landfill sites. The hydraulic conductivity of these clays is well characterized, but only under steady-state conditions. In reality, as with many situations in geology, landfills are dynamic environments and the lining materials are subject to temporal and spatial stress variations. These fluctuating stress conditions are likely to lead to deformation of the liner, which may in turn lead to changes in its hydraulic performance. To assess the behaviour of clay lining materials under operational stresses, samples of actual liner clays have been subjected to anisotropic loading in a triaxial cell while monitoring their hydraulic performance with a constant-rate-of-flow permeameter. The clays show marked changes in hydraulic conductivity during deformation, with increases and decreases being seen in different samples. However, the measured changes are probably not large enough to breach the minimum performance specification for landfill linings (current legislation calls for a maximum saturated hydraulic conductivity of 1×10^{-9} m s^{-1}). The fluctuations in hydraulic conductivity can be linked to the consolidation state and changes in the pore volume of the samples during deformation. The results and interpretations detailed here for three representative tests provide models of the hydrological behaviour of clays during deformation that may be applicable to poorly lithified mudrocks in other dynamic situations.

The nature of fluid flow through mudrocks is important in a number of fields, from oil recovery and development of ground-water supplies to radioactive and domestic waste disposal. Understanding of the flow behaviour is based on conventional permeability theory, which relies upon the material being homogeneous and the flow being steady state (Darcy 1856). However, in many geological situations the material being permeated is both anisotropic and inhomogeneous and may be undergoing active strain, leading to hydraulic conductivities that are dependent on the angular relationship between flow direction and material fabric (Wang & Anderson 1982). Deformation of poorly lithified mudrocks takes place principally by the development of shear zones, which leads to significant alteration of pore size and shape in the deformed zone (Maltman 1988). Therefore, because the nature of fluid flow in mudrocks is influenced by changes in pore geometry (Lin *et al.* 1986), the hydraulic performance is likely to change during deformation.

One situation in which mudrocks are deforming, with possible serious hydrological consequences, is in the linings of waste-disposal sites (Arch *et al.* 1996). Landfills are designed to isolate waste from the environment, and to do this they are constructed using a low hydraulic conductivity lining, currently specified to have a hydraulic conductivity of less than 1×10^{-9} m s^{-1}. This layer is designed to minimize the movement of water into, and leachate out of, the site. Although synthetic membranes, such as high-density polyethylene (HDPE), are used, the most common landfill-lining material is clay, because of its abundance, low cost, ease of emplacement and low natural hydraulic conductivity.

During and after the operational life of a site, the waste deposited in the landfill breaks down, either through biodegradation or through redox reactions (Edgers *et al.* 1992). This breakdown results in the reduction of volume of the fill, which leads to surface settlement and deformation of the capping layer (Edgers *et al.* 1992). Deformation of clays capping landfills has been shown to increase the hydraulic conductivity of the material by values in excess of an order of magnitude (Cressman *et al.* 1992). Similar movements can occur elsewhere in the liner as a result of side-slope failure or differential settlement at the base (Arch et al. 1996), so it seems reasonable to expect the hydraulic performance of the liner to change here also. This raises the long-term possibility of leachate leakage occurring from various parts of the landfill site. Leakage from shear zones or other flaws in the lining material could conceivably be responsible for the discrepancies between the hydraulic conductivities measured in the field and the lower values measured in the laboratory (Daniel 1984).

Consequently, to assess the significance of these effects, we have conducted an experimental programme to determine the changes in the hydraulic conductivity of clay landfill-lining material as it is exposed to operational stresses. The results and interpretations of three representative experiments are offered here, as they may provide models that are applicable to other situations where the fluid-flow behaviour of mudrocks in a dynamic environment is important. We begin by outlining some relevant background concepts and the testing procedures that we have employed.

Deformation and fluid flow in mudrocks

The response of weakly lithified mudrocks to an applied load depends principally on the interplay of four factors: deformation rate and hydraulic conductivity of the material, consolidation state, and the initial particle orientation. We briefly outline the role of each below. Deformation rate combined with the hydraulic conductivity of material determine whether the deformation will be drained or undrained, and hence whether or not excess fluid pressures will develop. In drained deformation the pore fluid can leave the material and fluid pressures remain in equilibrium. In undrained conditions the pore fluid is unable to dissipate; this leads to elevated pore pressures and hence lower material strengths (Lambe & Whitman 1969). Materials with low hydraulic conductivities, such as clays, tend to undergo undrained deformation even at low effective stresses and relatively low strain rates. Any fluctuations in hydraulic conductivity that occur during deformation can cause changes from drained to undrained deformation, and vice versa.

Consolidation state is the ratio of maximum past stress to the present load on the material (Lambe & Whitman 1969):

$$\text{OCR} = \frac{\sigma'_{\max}}{\sigma'_{\text{pres}}} \quad (1)$$

where OCR is the overconsolidation ratio, $\sigma'_{\max}$ is the maximum effective stress experienced and σ'_{pres} is the present effective stress.

Material with an OCR of unity is said to be normally consolidated, i.e. the present stress state is the maximum stress it has experienced. Material with an OCR of greater than unity is said to be overconsolidated. The stress and general volumetric response of clays of differing consolidation states to load is illustrated in Fig. 1. In overconsolidated clays shearing is accommodated by dilation of the material, giving the potential for increased hydraulic conductivity (Stephenson 1994). Many clays in the UK are naturally overconsolidated as a result of the removal of glacial ice, which produced high overburden pressures. Consequently, because the operational stresses placed upon the lining of the landfill are relatively low, the material will tend to be overconsolidated, meaning that any deformation is likely to result in an increase in hydraulic conductivity (Arch et al. 1996). Normally consolidated clays also behave in an

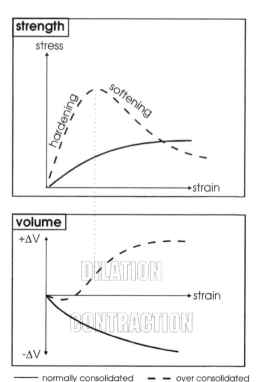

Fig. 1. Strength and volumetric response of overconsolidated and normally consolidated clay during deformation (after Mandl 1988).

overconsolidated manner. The process which leads to this change in behaviour is known as 'ageing' (Atkinson 1993); the main 'ageing' processes are compaction–vibration, creep, weathering and cementation (Atkinson 1993). This leads to the possibility of hydraulic conductivity increase in many landfill-lining clays.

The initial particle orientation influences deformation behaviour and is especially important in understanding how deforming mudrocks transmit fluid. Deformation of poorly lithified clay is accommodated principally by the sliding of particles rather than through grain breakage (Maltman 1987; Agar et al. 1989), and, because well-aligned mineral particles will slip relatively easily, the initial particle orientation closely influences shear behaviour and strength. The main types of particle orientation are shown in Fig. 2. The degree and pattern of flocculation is a function of load and, in low-stress situations, of soil water chemistry (Lambe & Whitman 1969). Although natural clays tend to show some degree of flocculation, mechanical working during emplacement of the liner will tend to convert any flocculated microfabric toward a more remoulded distribution. Exactly how the microstructure changes will depend on the conditions of placement, e.g. compactive effort, moisture content and the size of clods being compacted to form the liner (Wright et al. 1996).

Microstructural changes caused by stresses acting on the lining material during landfill operation can lead to changing values of hydraulic conductivity. For example, compactive stresses acting on a landfill lining with a flocculated particle distribution tend to cause overall collapse of the open pore framework and a bulk hydraulic conductivity decrease. If during compaction there is a marked increase in grain alignment an anisotropic hydraulic conductivity may develop, with an increased conductivity parallel to the collapsed, aligned, platy grains. If the tortuosity reduction is of a greater magnitude than the volume loss there can be a hydraulic conductivity increase despite overall contraction of the sample. Stresses caused by differential settling prompt the development of micro-shear zones, with reduced tortuosity and possibly with increased hydraulic conductivity parallel to the direction of fabric alignment within the zones (Arch & Maltman 1990). Despite the flow paths in clays being more inter-aggregate than inter-particulate in nature, particle alignments in shear zones have been proved to increase hydraulic conductivity (Arch & Maltman 1990). The results of the present work show that the factors outlined above interact closely in deforming mudrocks, sometimes in intricate ways.

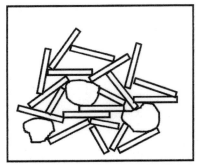
salt-water flocculation

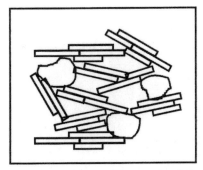
fresh-water flocculation

remoulded

Fig. 2. Common flocculation patterns of clays, highlighting the variation in porosity and tortuosity between the fresh-water flocculation pattern seen in non-worked clays and the remoulded distribution common in emplaced clay liners (after Lambe & Whitman 1969).

Sampling, equipment and methods

Fifteen clays have been tested in the experimental programme so far, and three representative results are discussed in some detail in the following section. All

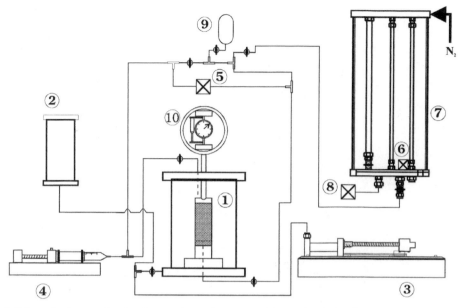

Fig. 3. Schematic diagram showing the experimental set-up used for testing hydraulic conductivity. The sample (shaded) is housed in a triaxial cell (1). The cell is filled and pressurized using an air–water cylinder (2). The cell pressure is controlled and measured by a GDS Instruments hydraulic actuator (3). Permeant is driven into the sample at a constant rate by the flow pump (4). The associated pressure head across the sample is measured by a differential pressure transducer (5), from which the hydraulic conductivity is calculated. The permeant discharged from the downstream end of the sample is measured by a differential pressure transducer (6) in the VCD (7). Back-pressure is applied to the system using a nitrogen (N_2) source attached to the VCD and measured using an absolute pressure transducer (8). The apparatus is filled from a reservoir (9). Axial stress is measured by displacement of a calibrated proving ring monitored by a displacement transducer (10).

three of these samples were of glacial origin, typical of the clays used in much of the UK as landfill liners. To account for some possible regional variation, two of the samples were from Wales and one from Scotland. In its natural state, the Scottish clay had a relatively large number of boulder-size clasts and a few clasts of pebble size. However, as the large clasts are routinely removed before emplacement of the liner, in practice it had fewer clasts than the Welsh boulder clay liner. In both cases, but especially with the Welsh sample, it was often necessary to cut more than one core to obtain a smooth-walled sample of appropriate dimensions. The laboratory study was conducted using small (38 mm diameter, 76 mm length) cylindrical samples of clay, cut either from the landfill liner directly or trimmed from samples of 100 mm diameter (U100) provided by the landfill operator. The sample size used during the research allowed us to collect small amounts of material from the liner, which could easily be refilled without compromising liner performance. Despite the diminutive dimensions of the cores they are large enough to incorporate a representative sample of the clay–silt fraction of the liner material.

The samples were then deformed in the laboratory using stresses similar in scale to those likely to be experienced in the landfill site. The values were based upon those reported by Peirce *et al.* (1986). Simultaneously, measurements were made of hydraulic conductivity and volume change in the sample. Hydraulic conductivity testing took place at a room temperature of 21°C.

The experimental equipment (Fig. 3) centres on a standard triaxial cell (Bishop & Henkel 1962), utilizing a hydraulic actuator manufactured by GDS Instruments Ltd to control and measure cell pressure. The cell is linked to a flow pump at the upstream end and a system of pressurized burettes, which make up the volume change device (VCD), at the downstream end. This provides a back-pressure, which ensures that any air trapped within the deionized water permeant is driven into solution. It also allows the pore fluid to be treated as incompressible; this means that all volumetric measurements will reflect pore volume change and not changes in fluid volume.

The saturated hydraulic conductivity is measured using the constant-rate-of-flow method (Aiban & Znidarcic 1989) involving a precise flow pump. This infuses fluid into the sample via a system of low-volume stainless steel tubes, which minimizes the compliance of the system. The flow is retarded by the sample and the resulting hydraulic head, measured by a differential pressure transducer, is a function of the flow rate and the hydraulic conductivity:

$$h = \frac{ql}{Ak} \qquad (2)$$

where h is the hydraulic head (L), l is the sample length (L), q is the flow rate ($L^3 T^{-1}$), A is area (L^2) and k is hydraulic conductivity (LT^{-1}) (after Darcy 1856).

The main advantages of the flow-pump technique are: (1) it is easier to control low flow rates than measure them; (2) the low flow rates allow the generation of low hydraulic heads and hence low hydraulic gradients across the sample; (3) testing time is reduced because there is no need to wait for a measurable volume of permeant to be passed through the sample (Olsen et al. 1985).

Using the flow pump and the VCD, the apparatus allows precise determination of volume changes within the deforming sample. The flow pump displaces permeant at a constant rate (9.999 ml min^{-1} to 0.001 ml h^{-1}) into the sample. Therefore, for a given interval of time, we can calculate the volume of permeant entry into the sample. Simultaneously, the downstream VCD measures the volume of permeant discharged from the sample. The VCD uses a differential pressure transducer to measure the difference in pressure between a column of air and a column of fluid linked to the downstream end of the sample. The changes in fluid height within the column can be converted, using the column dimensions and the density of water, to give the volume of fluid in the system. If more fluid is received by the VCD than was pumped into the sample over the same period of time, then the sample must be losing pore space (contracting) and discharging the fluid from the collapsed pores to the volume change device. Conversely, if there is less fluid received at the VCD than discharged from the flow pump, then the sample is dilating.

Results and discussions

Three representative examples of the 15 tests conducted are selected here and described in some detail, to illustrate the kind of data obtained and the interpretations we are able to make. The first test illustrates a hydraulic conductivity decrease with deformation, the second a transient increase, and the third a permanent increase in the capacity of the deformed clay to transmit fluid. Thus the three tests represent a range of hydraulic behaviour, and we believe that the processes we infer may be applicable to mudrocks deforming in other geological situations.

Welsh boulder clay

The sample was a core (U100) taken by the landfill operator from the basal lining of a new cell within a landfill site in the Ellesmere area, eastern Wales. The sample was tested under a confining pressure of 150 kPa and a back-pressure of 100 kPa (although the back-pressure used was not high enough to ensure saturation of the sample, the volume of fluid pumped into and received from the sample was monitored during the static testing; the inflow and outflow volumes were the same during this period, indicating no change in the volume of permeant within the sample: this is inferred to mean that the sample is at or near saturation), reproducing the effective stress conditions expected by the landfill operator. The flow rate during testing was 0.002 ml min^{-1}. Under these static test conditions the sample gave a saturated hydraulic conductivity of 7.282×10^{-10} m s^{-1}. The sample was then subjected to an axial strain rate of 9.02×10^{-7} s^{-1} until failure, marked by the sudden drop in axial stress at a deviator stress of 110 kPa after 8% strain (Fig. 4). During deformation the hydraulic conductivity of the sample decreased, though in a non-linear way, especially before failure. During strain hardening (at 2–3.5% strain), there was a period of hydraulic conductivity increase from 6.075×10^{-10} s^{-1} to 6.2×10^{-10} s^{-1}, coinciding with a small change in stress/strain response for the sample (Fig. 4). During this time there was a reduction in the rate of fluid flow to the VCD, indicating a drop in the rate of fluid leaving the sample (Fig. 5). The most striking feature of Figs 4 and 5 is the rapid drop in hydraulic conductivity at the point of sample failure, coincident with the volume discharged to the VCD increasing rapidly. We interpret this as a pulsed release of stored fluid, but it is important to note that conventional methods of determining hydraulic conductivity would interpret this outpouring of permeant simply to be an increase in hydraulic conductivity. The volume-change measurements allowed by our experimental arrangement indicate that the latter would be an erroneous conclusion.

The small initial increase in hydraulic conductivity seen in the Welsh boulder clay may be a result of the transition from sample consolidation (where excess pore fluid pressure caused by loading is dissipated and the sample loses volume) to the initiation of shearing (resulting in pore dilation). This inference is supported by the reduction in the pore fluid discharge during the same interval (up to 8% strain) as that indicating a decrease in pore collapse during deformation. The rapid decrease in hydraulic conductivity of the sample during failure indicates a collapse of porosity in excess of any decrease in tortuosity during failure.

Scottish boulder clay

The sample of Scottish boulder clay was cored from the lining of a landfill outside Glasgow using a 38 mm coring bit. During testing the experiment conditions were a confining pressure

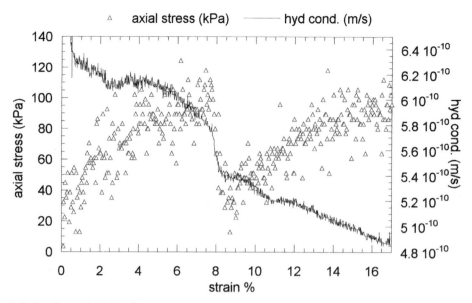

Fig. 4. Hydraulic conductivity of the Welsh boulder clay during deformation. (Note the drop in hydraulic conductivity at the point of failure.)

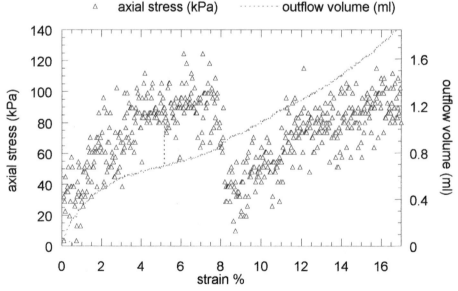

Fig. 5. Discharged volume from the Welsh boulder clay during deformation. (Note the increase in discharged volume during failure.)

of 400 kPa, a back-pressure of 200 kPa and a flow rate of 0.002 ml min^{-1}. These conditions gave a higher effective stress than for the Welsh boulder clay. The conditions were chosen because the landfill was to receive contaminated soils, which have a higher density and hence produce a greater load than the domestic waste deposited at the Ellesmere site. Under these static conditions the sample had a saturated hydraulic conductivity of 1.06×10^{-10} m s^{-1}. The sample was then deformed at a strain rate of 2.74×10^{-6} s^{-1}, until failure at an axial stress of 200 kPa after 17% strain. The initial loss of sample strength seen at 0–4% strain was a result

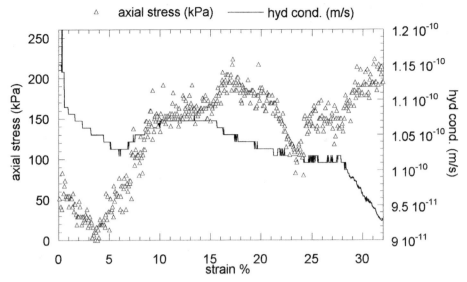

Fig. 6. Hydraulic conductivity of the Scottish boulder clay during deformation, illustrating a transient increase in hydraulic conductivity during strain hardening.

of the establishment of the hydraulic head, which causes a reduction in the effective normal stress acting on the sample, thus allowing easier sliding of the grains, and hence a loss of strength.

During deformation the hydraulic conductivity of the material changed in a complex way (Fig. 6). Up to 4% strain there was an apparent decrease in the hydraulic conductivity. Again, this was a result of the establishment of the hydraulic head. This was followed by a period of strain hardening and dilation marked in Fig. 7 by a peak into the dilational field of the interval volume diagram. Interval volume change is calculated by subtracting the volume received at the VCD from the volume pumped into the sample over a fixed time period (in this case every 180 s). Positive numbers therefore indicate dilation and negative numbers indicate contraction. During the dilation there was continued hydraulic conductivity reduction. At 6.5% strain the dilation event ceased and the hydraulic conductivity increased from 1.025×10^{-10} to 1.075×10^{-10} m s^{-1}, a 4.9% increase in hydraulic conductivity. The sample's hydraulic conductivity started to decrease again at the second phase of strain hardening, which occurred at 16% strain. The decrease continued until failure was complete at 23% strain, where the hydraulic conductivity remained relatively constant through the final phase of strain hardening until 28% strain. At this point the hydraulic conductivity rapidly dropped (Fig. 6).

The dilation event arises from the development of shear zones in the deforming sample. These zones act as storage sites for fluid, hence the continued drop in hydraulic conductivity. However, with continued deformation the zones start to act as conduits for flow. This is probably caused by an increase in interconnectivity between the dilatant pores, which, combined with grain alignment, produces a decrease in tortuosity. Once these sites are connected the overpressured fluid can escape, leading to the contractional event seen from 6% strain. Interconnectivity of pores is maintained and this leads to the development of enhanced fluid flow pathways, which increases the hydraulic conductivity of the material. With continued deformation the shear zones lock up and collapse, causing the final strain hardening event and the accompanying decrease in hydraulic conductivity.

Welsh boulder clay landfill bund

The second sample of Welsh boulder clay was cut from a U100 core supplied by the landfill operator, from a bund dividing the cell under construction from the rest of the site. Under steady-state isotropic conditions with an effective confining pressure of 40 kPa (the estimated overburden pressure acting upon the bund during operation), the sample had a saturated hydraulic conductivity of 4.06×10^{-9} s^{-1}. The

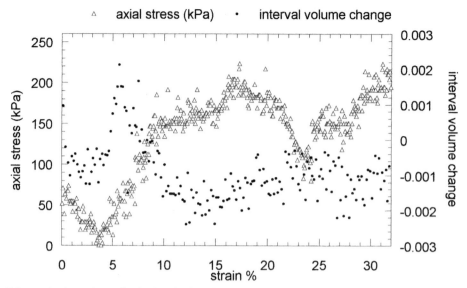

Fig. 7. Interval volume change for the Scottish boulder clay during deformation, with positive values during strain hardening indicating dilation.

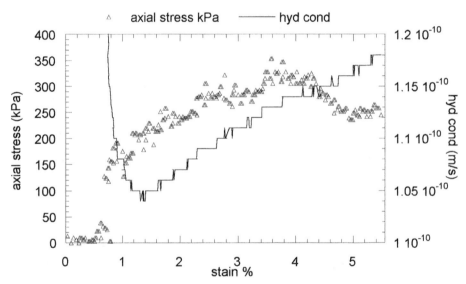

Fig. 8. Hydraulic conductivity of the Welsh landfill bund material during deformation, showing hydraulic conductivity increasing with continued deformation.

sample was then deformed under a confining stress of 380 kPa and a back-pressure of 146 kPa, and was infused with permeant at a rate of 0.002 ml min^{-1}. The sample failed at 3.75% strain under an axial load of 330 kPa (Fig. 8). Synchronous with the deformation was an increase in hydraulic conductivity from 1.042×10^{-10} to 1.18×10^{-10} s^{-1} (Fig. 8). The VCD records an initial period of rapid dewatering from the sample until 1% strain followed by a fairly constant rate of outflow, the transition between the two outflow rates occurring at the end of the period of extremely rapid strain hardening (Fig. 9).

The sample shows a relatively large increase in hydraulic conductivity with deformation after an

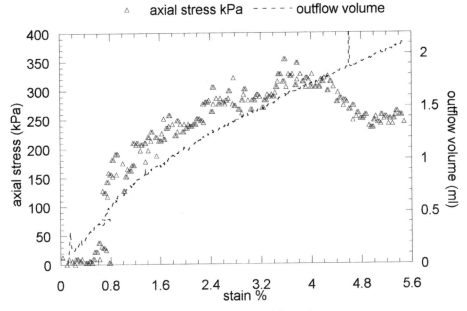

Fig. 9. Outflow volume of Welsh landfill bund material during deformation.

initial decrease, and does not exhibit the general increase in hydraulic conductivity with strain shown by the other samples. The hydraulic conductivity is less than under static conditions, but this is to be expected as the sample is under a greater stress, which reduces the effective pore radius. The initial decrease in hydraulic conductivity is due to the material consolidating in response to the first increments of loading. The consolidation episode is responsible for the increase in volume discharged to the VCD. With increasing strain the deformation is accommodated by the development of shear zones in the sample. The reorientation of grains during shear reduces the tortuosity of the sample, and this results in the observed increase in hydraulic conductivity. There is very little dilation of the sample recorded in the VCD response. The behaviour of the sample is due to the microfabric of the sample having a structure resembling the remoulded shear. Therefore, deformation would be accommodated by the sliding of preferentially oriented zones of particles. The small volume losses seen during testing suggest that tortuosity reduction would be the major change in the microfabric of the sample, which leads to the observed increase in hydraulic conductivity. The development of a remoulded microfabric is likely in the material used to construct a bund, as it is subject to large amounts of rolling and compaction to produce the final structure.

Conclusions

The results indicate that the performance of landfill-lining clays changes with deformation and therefore will vary during the life of the site. The pattern and magnitude of the change is linked to the chemical and physical properties of the material and its deformation characteristics. The interplay between pore pressure, hydraulic conductivity and deformation is complicated, but definable in each sample tested. In two of the tests there is an overall decrease in the hydraulic conductivity of the material with deformation. This suggests that the emplacement of the liner leads to a high degree of remoulding within the sample, which results in the material behaving in a normally consolidated manner. The instances of hydraulic conductivity increase in these materials are restricted to short periods of the deformation history and are of a smaller scale than the overall decrease in hydraulic conductivity. However, given the lower strain rates in nature and the possibility of a non-uniform deformation rate, the period of decreased liner performance may in reality last for a considerable length of time.

For areas of landfill material that are overcompacted, like the bund material, there is a tendency for a large hydraulic conductivity increase. In marginal cases, where a liner's performance is close to the minimum specification and there are likely to be large numbers of deforming areas, the danger of leachate leakage is very real. For the majority of cases, however, liners are constructed from materials that surpass the required tolerances by a considerable margin, so that the changes in performance of the magnitude seen in the laboratory are unlikely to lead to any leakage of liquid pollutants from the site.

The results highlight the importance of strain, particle alignment and pore geometry on the hydrological behaviour of poorly lithified mudrocks, and, especially, the critical role of consolidation state. Such factors should be considered when analysing the fluid-flow behaviour of mudrocks in any geological situation where deformation is likely.

The authors would like to thank A. Bolton for numerous discussions on the hydrology of deforming clays, and B. Clennell for writing the computer code used to run the experimental equipment.

References

AGAR, S. M., PRIOR, D. J. & BEHRMANN, J. H. 1989. Back-scattered electron imagery of the tectonic fabrics of some fine-grained sediments: implications for fabric nomenclature and deformation processes. *Geology*, 17, 901–904.

AIBAN, S. A. & ZNIDARCIC, D. 1989. Evaluation of the flow pump and constant head techniques for permeability measurements. *Géotechnique*, 39, 655–666.

ARCH, J. & MALTMAN, A. J. 1990. Anisotropic permeability and tortuosity in deformed wet sediments. *Journal of Geophysical Research*, 95, 9035–9045.

——, STEPHENSON, E. & MALTMAN, A. J. 1996. Factors affecting the containment properties of natural clays. *In*: BENTLEY, S. P. (ed.) *Engineering Geology of Waste Disposal*. Geological Society, London, Engineering Geology Special Publication, 11, 259–265.

ATKINSON, J. H. 1993. *Introduction to the Mechanics of Soils and Foundations*. McGraw–Hill, London.

BISHOP, A. W. & HENKEL, D. J. 1962. *The Measurement of Soil Properties in the Triaxial Test*, 2nd edn. Edward Arnold, London.

CRESSMAN, G. C., CHENG, S. C., MARTIN, J. P. & BREDARIOL, A. S. 1992. Effects of subsidence distortion on integrity of landfill caps. *In*: USMAN, M. A. & ACAR, Y. D. (eds) *Environmental Geotechnology*. Balkema, Rotterdam, 229–235.

DANIEL, D. E. 1984. Predicting hydraulic conductivity of clay liners. *Journal of Geotechnical Engineering, American Society of Civil Engineers*, 110, 285–300.

DARCY, H. 1856. *Les fontaines publiques de la ville de Dijon*. Dalmont, Paris, 590–594.

EDGERS, L., NOBLE, J. J. & WILLIAMS, E. 1992. A biologic model for long term settlement. *In*: USMAN, M. A. & ACAR, Y. D. (eds) *Environmental Geotechnology*. Balkema, Rotterdam, 177–184.

LAMBE, T. W. & WHITMAN, R. V. 1969. *Soil Mechanics*. Wiley, New York.

LIN, C., PIRIE, G. & TRIMMER, D. A. 1986. Low permeability rocks: laboratory measurement and three dimensional microstructure analysis. *Journal of Geophysical Research*, 91, 2173–2181.

MALTMAN, A. J. 1987. Shear zones in argillaceous sediments — an experimental study. *In*: JONES, M. E. & PRESTON, R. M. F. (eds) *Deformation of Sediments and Sedimentary Rocks*. Geological Society, London, Special Publications, 29, 77–87.

—— 1988. The importance of shear zones in naturally deforming wet sediments. *Tectonophysics*, 145, 163–175.

MANDL, G. 1988. *Mechanics of Tectonic Faulting — Models and Basic Concepts*. Elsevier, Amsterdam.

OLSEN, H. W., NICHOLS, R. W. & RICE, T. L. 1985. Low-gradient permeability measurements in a triaxial system. *Géotechnique*, 18, 145–157.

PEIRCE, J. J., SALLFORS, G. & MURRAY, L. 1986. Overburden pressures exerted on clay liners. *Journal of Environmental Engineering, American Society of Civil Engineers*, 112, 280–291.

STEPHENSON, E. L. 1994. *A laboratory investigation of fluid flow in deforming sediments*. PhD thesis, University of Wales, Aberystwyth.

WANG, H. F. & ANDERSON, M. P. 1982. *Introduction to Groundwater Modelling, Finite Difference and Finite Element Models*. Freeman, Oxford.

WRIGHT, S. P., WALDEN, P. J., SANGHA, C. M. & LANGDON, N. J. 1996. Observations on soil permeability, moulding moisture content and dry density relationships. *Quarterly Journal of Engineering Geology*, 29, 249–255.

Gas transport properties of clays and mudrocks

J. F. HARRINGTON & S. T. HORSEMAN

Fluid Processes Group, British Geological Survey, Nottingham NG12 5GG, UK

Abstract: Controlled flow rate gas injection experiments have been performed on clay and mudrock samples using helium as a permeant. By simultaneously applying a confining stress and back-pressure, specimens were isotropically consolidated and fully water saturated under predetermined effective stress conditions, before injecting gas at a very slow rate using a syringe pump. Ingoing and outgoing gas fluxes were monitored. All tests exhibit a conspicuous threshold pressure for gas breakthrough. All tests showed a post-peak negative transient leading to steady-state gas flow. On the basis of a stepped history of flow rate, the flow law was shown to be nonlinear (i.e. non-Darcian). With the injection pump stationary (i.e. zero flow rate), gas pressure declined with time to a finite value. No gas flow was ever detected at excess gas pressures less than this lower threshold. When gas flow was re-established, the threshold for gas breakthrough was found to be significantly lower than in the virgin clay. There is strong evidence to suggest that the capillarity restrictions on gas penetration of the intergranular pores of saturated clays and mudrocks are of such a magnitude that normal two-phase flow is impossible. Gas therefore does not occupy, or flow through, the intergranular porosity of the clay matrix. In the absence of pressure-induced cracks, water-saturated clays and mudrocks are totally impermeable to gas. The measured gas permeability of a clay-rich medium is a dependent variable rather than a material property, as it depends on the number of pressure-induced pathways in the plane normal to the flow, together with the width and aperture distributions of these pathways. The experiments suggest that the flow pathways open under high gas pressure conditions and partially close if gas pressure falls, thus providing a possible explanation of the nonlinearity of the flow law. Reliance on conventional two-phase flow theory is inadvisable when attempting to quantify gas transport in initially water-saturated clay-rich materials.

Compact clays and mudrocks should represent very significant barriers to fluid migration. However, if the forces that drive migration rise to the point where they exceed the resistance of these barriers, then fluids will move by exploiting 'preferential pathways'. In many cases, the incipient pathways of fluid flow are already present as faults or as interconnecting networks of smaller fractures or fissures. In special circumstances, very high fluid pressures can actually create the pathways of fluid migration through clay formations. Hedberg (1974), in a paper on the role of gases in the overpressurization of oilfield shales, provided us with a translation of a very important passage from a paper by Tissot & Pelet (1971) which states: 'The displacement of an oil or gas phase from the centre of a finely-divided argillaceous matrix goes against the laws of capillarity and is in principle impossible. This barrier can, however, be broken in one way. The pressure within the fluids formed in the pores of the source-rock increases constantly as the products of the evolution of the kerogene are formed. If this pressure comes to exceed the mechanical resistance of the rock, microfissures will be produced which are many orders of size greater than the natural (pore) channels of the rock, and will permit the escape of an oil or gas phase, until the pressure has fallen below the threshold which allows the fissures to be filled and a new cycle commences.'

Tissot and Pelet's comments sum up modern concepts of gas-phase transport in compact clay-rich media in a remarkably succinct manner. In an initially water-saturated clay with extremely narrow interparticle spaces, the capillary threshold for gas entry is simply too large for the gas to penetrate and desaturate the clay matrix. However, when large gas pressures relative to the boundary stress are applied to a clay, some sort of fracturing process occurs. Given a suitable exit route, the gas then traverses the clay by moving entirely within pressure-induced pathways. Although this type of behaviour has been widely discussed by oil industry researchers in the context of the primary migration of hydrocarbons from source-rock shales (Duppenbecker et al. 1991), our experimental study suggests that it is fundamental to all compact clay-rich media.

The mathematical theories of two-phase and multi-phase flow in porous media (Aziz & Settari 1979; de Marsily 1986) are based on the semi-empirical generalization of Darcy's law. These theories stem from oil industry interest in the movement of gaseous and liquid hydrocarbons in geological media and are widely applied to the problems and rock-types of concern to that industry. In developing two-phase flow theory, it is assumed that Darcy's law is valid for each fluid phase considered separately, as if it occupied a certain proportion of the porous medium. The Darcy velocity of the gas, q_g (m s^{-1}), is given by

$$q_g = -k \frac{k_{rg}}{\eta_g}(\nabla p_g + \rho_g g \nabla z) \qquad (1)$$

and that of the water, q_w, by

$$q_w = -k \frac{k_{rw}}{\eta_w}(\nabla p_w + \rho_w g \nabla z) \qquad (2)$$

where k (m^2) is the intrinsic (or absolute) permeability of the porous medium, η (Pa s) is viscosity, ρ (Mg m^{-3}) is density, p (Pa) is pressure, g is the acceleration due to gravity, z is the elevation and the subscripts g and w denote gas and water, respectively. The relative permeabilities to gas and water, k_{rg} and k_{rw}, are dimensionless variables, normally in the range 0–1, which relate the effective permeability of the porous medium to a particular phase to the intrinsic permeability, k, when the medium is partly saturated by the second phase,

$$k_{rg} = \frac{k_g}{k} \quad \text{and} \quad k_{rw} = \frac{k_w}{k}. \qquad (3)$$

The intrinsic permeability is considered to be a property of the medium alone and can be calculated from hydraulic conductivity, K (m s^{-1}), using

$$k = \frac{K \eta_w}{\rho_w g} \qquad (4)$$

The phase saturations are defined by

$$S_g = \frac{V_g}{V_p} \quad \text{and} \quad S_w = \frac{V_w}{V_p} \qquad (5)$$

where V_g and V_w are the volume of gas and water, respectively, and V_p is the total pore volume in a representative elementary volume (REV) of the porous medium. If n is the porosity of the medium, then the product nS_g can thus be thought of as the effective porosity of that portion of the partitioned pore space occupied by the gas phase. The product nS_w is known in soil physics as the volumetric water content, θ_w. Within the REV, $S_g + S_w = 1$, even when saturation is changing with position and time. The relative permeability to gas, k_{rg}, is normally considered to exhibit a first-order dependence on the gas saturation.

Gas–water meniscus

The difference in pressure, p_c, between gas and water, in equilibrium and separated by a stable interface, is known as matric suction in unsaturated soil mechanics (Fredlund & Rahardjo 1993) and capillary pressure in oil industry studies (Aziz & Settari 1979). It is usually quantified by the Young–Laplace relationship

$$p_c = p_g - p_w = \gamma_{wg}\left(\frac{1}{r_1} + \frac{1}{r_2}\right) \qquad (6)$$

where r_1 and r_2 are the principal radii of curvature of the meniscus (positive if they point into the gas phase) and γ_{wg} is the interfacial tension between water and gas. Although there are many models for pore geometry, we can think provisionally in terms of a system of interconnected tubular capillaries representing the flow channels. Taking the gas–water meniscus as hemispherical, the radius of curvature of this meniscus is

$$r_1 = r_2 = \frac{a}{\cos \phi} \qquad (7)$$

where a is the capillary radius and ϕ is the contact angle. The criterion for the passage of gas into a capillary channel becomes

$$p_g > p_w + \frac{2\gamma_{wg}}{a} \cos \phi. \qquad (8)$$

The porous medium is assumed to have a distribution of capillary sizes, which determines the relationship between p_c and S_w (saturation function) or p_c and θ_w (water-retention function). If the gas pressure exceeds the local water pressure by an amount sufficient to overcome surface tensional forces, then the gas can enter the flow channel.

Water-retention function

The water-retention function (or moisture characteristic) is a plot of matric suction against volumetric water content. For ultra-fine-grained materials with high specific surface, such as clays, the retention function may not be solely determined by capillarity. A proportion of the pore water can be adsorbed or bonded to mineral surfaces, and this makes it very difficult to displace (Newman 1987). The retention function can be evaluated for an adsorption history (water content increasing) or a desorption history (water content decreasing). The desorption curve is of specific interest to the gas transport problem and can be evaluated using a number of well-known techniques (Marshall & Holmes 1979). One method uses the pressure-plate apparatus. A sample is placed on a cellulose acetate membrane supported by a water-saturated porous plate. This plate is linked to an external apparatus that maintains a fixed pore-water pressure, p_{wo}, in the plate. The sample is enclosed in a pressure vessel that allows air pressure, p_{gi}, acting on the upper surfaces of the sample to be raised with respect to the controlled water pressure. During a test, air pressure is increased in fixed steps and then held constant until drainage of water out of the sample has ceased and the system is in equilibrium. On the basis that the matric suction is simply the difference between the applied air pressure and the downstream water pressure,

$$p_c = p_{gi} - p_{wo} \qquad (9)$$

a graph can be constructed of p_c against θ_w over a range of air pressures. Interpretation of test results is straightforward for relatively incompressible, framework-supported materials such as sand. However, many clays exhibit significantly large volume changes when subject to varying gas pressures. Up to the point of gas entry, the applied gas pressure exerts a total stress on the sample, leading to consolidation and dewatering.

Air entry value

Gas entry into a clay-rich material is believed to occur at some critical value of matric suction, known as the air entry value (AEV). An indication of the likely magnitude of the AEV can be obtained from the clay shrinkage response (Delage & Graham 1995). It is widely recognized that, when an initially saturated clay is slowly dried by exposure to a low-humidity atmosphere, it undergoes a gradual reduction in volume. If both the volume and the mass of the sample are monitored with time and a graph is plotted of water content against sample volume per unit mass of dry solids, it is clear that the initial changes in volume are entirely explained by loss of water. No air enters the clay and the data points on the graph plot precisely on the saturation line. If the relative humidity is further reduced, the plotted data points start to move away from this line, indicating that air entry has taken place. This occurs at a water content somewhat larger than the shrinkage limit (geotechnical index property). Provided that the humidity conditions are controlled and the sample is allowed to fully equilibrate at each stage, the critical suction corresponding to the AEV can be calculated using the thermodynamic relationship

$$p_c = -\frac{RT}{v_{mw}} \ln\left(\frac{p_v}{p_{va}}\right) \qquad (10)$$

where p_v is the controlled vapour pressure of the pore water in the clay sample, p_{va} is the vapour pressure of bulk pore water when at atmospheric pressure, R is the gas constant, T is temperature and v_{mw} is the partial molar volume of the water. The Kelvin relationship of physical chemistry can be obtained by combining equations (6) and (10).

This response of a clay during shrinkage is illustrated by the experimental study of Croney & Coleman (1953). Shrinkage tests on London Clay and Gault clay show departures from the saturation line at gravimetric water contents of 22% and 17%, respectively. Using the supplied water-retention curves, the AEV of the London Clay is around 1 MPa and that of Gault clay around 6.3 MPa. Black (1962) showed that the higher the plasticity of a clay, the higher the critical suction at which desaturation occurs. The higher AEV of the Gault clay can therefore be explained by the presence of smectite clay minerals, which results in a material with high plasticity.

When we insert a gas pressure equal to atmospheric pressure (c. 0.1 MPa) and a matric suction equivalent to the AEV for Gault clay into equation (6), we find that the pore-water pressure in the clay at the time of air entry was around −6.2 MPa (i.e. negative) in Croney & Coleman's experiments. This raises the important question of the physical significance of a negative pore-water pressure. It seems probable that the water is in a state of tension. Inspection of the phase diagram for water is not very informative, as tensile forces are not considered in physical chemistry. Nevertheless, there is little doubt that pore water in tension is a thermodynamically metastable phase, which is prevented from vaporizing by the extreme narrowness of the interparticle spaces.

There is no unique interpretation of the mechanism of gas entry during shrinkage. One concept is that the reduction in volume of the clay associated with desiccation in the region of the shrinkage limit is opposed by the development of large forces of repulsion between neighbouring clay particles. Although these might be Hertzian-type mineral–mineral (M–M) contact forces, given the strength of bonding of water molecules to clay surfaces (Newman 1987), it seems more probable that the repulsive forces are transmitted by molecularly thick diameter water films between clay particles. As further drying cannot easily be accommodated by volume change, the remainder of the pore water (outside the regions of close approach) must develop a very large internal tension tending to pull the clay particles together. We equate this tension with the matric suction. The possible mechanisms of gas entry then become as follows: (a) the tensile stress exceeds the local strength of the water films, which leads to film rupture and air penetration of the newly formed cracks, or (b) the pore-water suction exceeds the maximum sustainable capillary pressure drop across the bordering gas–water meniscus, and this leads to the inward movement of the air.

The macroscopic effects of clay desiccation are well known. When the critical value of suction is exceeded, a clay ruptures to form an array of shrinkage cracks. These often exhibit the polygonal arrangement typical of two-dimensional extensile rupture. Vaporization of the thermodynamically metastable liquid phase can be postulated as a possible mechanism for the nucleation of these cracks. Figure 1 shows a polygonal array of shrinkage cracks in the upstream face of a specimen of pre-compacted bentonite after gas testing and drying. The characteristic repeat spacing of the cracks is around 6 mm. Extreme desiccation has produced very dilated cracks with apertures in the range 0.05–0.5 mm. These cracks represent the pathways of air entry into bentonite when subject to very high matric suction.

The principle of axis-translation is widely applied in laboratory studies of unsaturated particulate systems (Fredlund & Rahardjo 1993; Delage & Graham 1995). It states that the matric suction depends on the difference between the gas pressure and the measurable water pressure and is therefore independent of the absolute magnitude of either quantity. Two clays can therefore have identical equilibrium suctions if the first has a negative pore-water pressure (i.e. thin films in a state of tension) and is exposed to gas at atmospheric pressure, but the second has a pore-water pressure equal to

Fig. 1. Low-magnification view ($\times 3.5$) of the upstream face of a pre-compacted Mx80 bentonite after gas testing and oven drying at 105°C for 24 h, showing a typical polygonal arrangement of extensile shrinkage cracks. The characteristic repeat spacing is c. 6 mm. The shrinkage cracks are dilated by extreme desiccation of the bentonite and have typical apertures in the range 0.05–0.5 mm. These cracks are the pathways of air entry into bentonite when subject to very high matric suction.

atmospheric pressure and is exposed to gas at a pressure greater than atmospheric. Axis-translation suggests some degree of commonality between the mechanisms of gas entry in an initially water-saturated clay or mudrock subject to compressive boundary stresses and positive pore-water pressures (i.e. conditions at some depth below ground surface) and those that lead to the development of gas-filled cracks in a clay subject to desiccation, shrinkage and negative pore-water pressures.

Experiments

The controlled flow rate permeameter used in the experimental study (Fig. 2) comprises five main components: (a) a specimen assembly; (b) a pressure vessel together with its associated pressure control equipment; (c) a fluid injection system; (d) a back-pressure system; (e) a microcomputer-based data acquisition system. Conceptually, the equipment is similar to the pressure-plate apparatus but with two important points of difference. Unlike the pressure-plate apparatus, the specimen in our experiments is sheathed and subject to an isotropic confining stress, σ, which is held constant. Second, there is no cellulose acetate membrane at the downstream end of the specimen so that gas is able to flow in the axial direction. The clay is equilibrated by back-pressuring with water or synthetic pore solution at fixed pressure. The pressure vessel (Fig. 3) comprises a custom-built single closure vessel, manufactured from 316 stainless steel and rated to 70 MPa. The 4.9 cm diameter cylindrical clay specimen is sandwiched between two tapered end-caps, each with a sintered stainless steel porous disc, and jacketed in a thin-walled (0.2 mm) copper sheath to exclude confining fluid and prevent diffusional losses of gas. Tapered locking rings ensure a very gas-tight seal. The combination of sizes guarantees an interference fit between the clay and the copper. Heat-shrinkable Teflon tubing has been used in some more recent tests, and provides greater flexibility than copper. The injection end-cap has a central inflow duct and a circular groove cut into its load-bearing surface and linked to an outflow duct; this arrangement allows the gas to sweep radially through the porous disc during flushing operations.

The volumetric flow rate of the injected fluid and the pressure of the downstream fluid are controlled using a pair of ISCO-500, Series D, syringe pumps, operating from a single digital control unit. A pressure transducer monitors the outgoing pressure and provides a feedback signal to the microprocessor when the pump is set in pressure control mode. Piston motion gives a

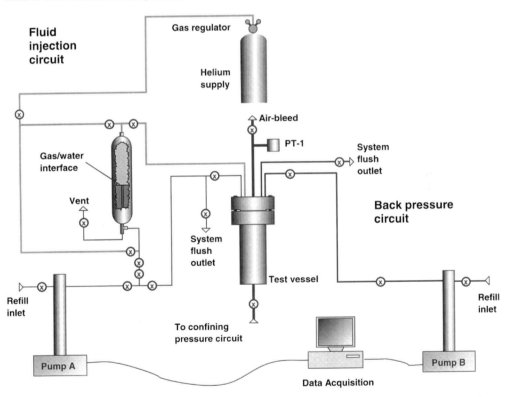

Fig. 2. Schematic diagram of the controlled flow rate gas permeameter. The five main components are a specimen assembly, a pressure vessel together with its associated pressure control equipment, a fluid injection system, a back-pressure system, and a microcomputer-based data acquisition system.

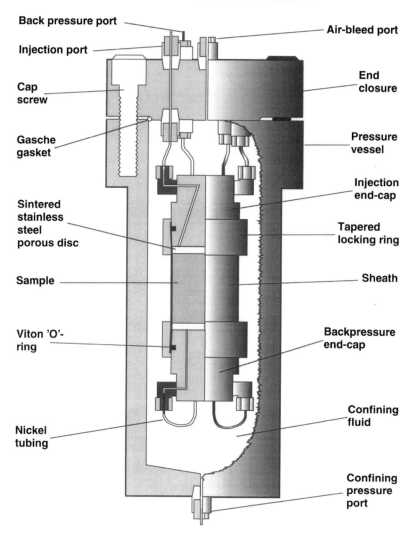

Fig. 3. Schematic diagram of the pressure vessel and specimen assembly. The pressure vessel is a custom-built stainless steel vessel rated to 70 MPa. The cylindrical clay specimen is sandwiched between two tapered end-caps, each with a sintered stainless steel porous disc, and jacketed in a copper or Teflon sheath to exclude confining fluid.

direct measure of the volumetric flux of fluid. These pumps can also be set to constant pumping rate mode, whereby the syringe piston is advanced at constant velocity. In gas injection experiments we use helium as a permeant. Given the possibility that gas might leak past the seal of the syringe, we opt to pump water and develop a constant helium flow rate by displacing it from a pre-charged vessel. This also ensures that the helium is water saturated and cannot therefore cause desiccation of the clay. Experiments are performed in an air-conditioned chamber at a temperature of $20 \pm 0.3°C$.

The ISCO pump controller has an RS232 serial port, which allows volume, flow rate and pressure data from the pumps to be transmitted to the equivalent port of a personal computer. A program written in QBASIC prompts the pump controller to transmit the data to the computer at preset time intervals. Typical acquisition rate is one scan per 30 min. Confining pressure is monitored using a pressure transmitter linked to a Druck DPI-203 digital pressure indicator. All pressure sensors and associated electronics are calibrated against a Druck PTX-610 pressure transmitter, calibrated by the manufacturer to ± 6.9 kPa. Least-squares linear regression of these data provides the slopes and intercepts necessary for processing the logged data. All quoted pressures are gauge pressures (relative to atmospheric pressure).

Specimens

Undisturbed samples of the moderately plastic and overconsolidated Oligo-Miocene Boom Clay were taken, under the supervision of the Belgian company SCK/CEN, by jacking a series of stainless steel sampling tubes with sharpened cutting edges into the exposed floor of the test gallery, at a location 224 m below surface in the HADES underground laboratory at Mol in Belgium. Blocks of pre-compacted bentonite were manufactured by Clay Technology AB (Lund, Sweden) by rapidly compressing Volclay Mx80 bentonite powder in a mould under a one-dimensionally applied stress (Johannesson et al. 1995). Mx80 is a trade name of Volclay Ltd, a subsidiary of the American Colloid Company. The material is fine-grained sodium bentonite from Wyoming containing around 90% montmorillonite. The bentonite blocks were cut into rectangular prisms with a band saw. Cylindrical specimens were prepared with the aid of a 4.9 cm i.d. stainless steel ring-former with a sharp leading edge. The clay was carefully pared away around the periphery using a scalpel, allowing the former to slide downward under a light, manually applied pressure. The upper and lower surfaces were trimmed and finished-off by a scraping action, to leave the ends perfectly flat and parallel. The clay was extruded from the former into the sheath. The porous discs were water saturated before the specimen assembly was made up.

Table 1 shows the basic physical properties of the test specimens based on pre-test measurements of water content and bulk density. A specific gravity of 2.77 was assumed for bentonite particles. Water content was determined by oven-drying at 105°C for a period in excess of 24 h. Bentonite blocks were slightly 'desaturated' in the as-supplied condition and physical properties are corrected for the likely effect on saturation of applying the confining stress. The bentonite specimens were divided into two batches, identifed as HS for 'high swelling' with swelling pressures, Π_{sw}, in the range 14–17 MPa and MS for 'medium swelling' with swelling pressures in the range 6–10 MPa. Intrinsic permeabilities for Boom Clay were measured using a multi-stage controlled flow rate technique with synthetic pore solution as a permeant. Values for Mx80 bentonite were calculated from the trends given by Börgesson et al. (1996).

Procedures

Specimens were water saturated ($S_w > 0.99$) and equilibrated under stress with a back-pressure applied at both ends. Test conditions are given in Table 2. Net flows were monitored to establish the point of equilibration. The intrinsic permeability of the Boom Clay specimens was then measured by the controlled flow rate method. After pore pressure equilibration, helium was admitted into the upper part of the gas–

Table 1. *Details and pre-test physical properties of the test specimens*

Material/ Details*	Test number	Water content (wt%)	Bulk density (Mg m^{-3})	Dry density (Mg m^{-3})	Void ratio	Intrinsic permeability (m^2 × 10^{20})
Boom Clay						
Norm.	T1S1	25.3	2.001	1.598	0.677	10
Norm.	T2S1	23.1	2.032	1.650	0.617	9
Norm.	T2S2	23.9	2.008	1.621	0.633	17
Norm.	T3S3	23.2	2.043	1.659	0.624	–
Averages		*23.9*	*2.021*	*1.632*	*0.638*	*12*
Par.	T2S3	23.1	1.994	1.620	0.596	46
Par.	T3S1	23.4	2.048	1.660	0.635	32
Par.	T3S2	22.7	2.060	1.678	0.617	48
Averages		*23.1*	*2.034*	*1.653*	*0.616*	*42*
Bentonite						
HS	Mx80-1	23.5	2.072	1.678	0.651	0.6
HS	Mx80-2	23.5	2.072	1.678	0.651	0.6
HS	Mx80-3	23.1	2.079	1.689	0.640	0.5
HS	Mx80-4	23.8	2.067	1.669	0.659	0.6
Averages		*23.5*	*2.073*	*1.678*	*0.650*	*0.6*
MS	Mx80-5	28.2	1.994	1.555	0.781	1.0
MS	Mx80-6	26.2	2.026	1.605	0.726	0.9
MS	Mx80-7	27.5	2.004	1.572	0.762	1.0
Averages		*27.3*	*2.008*	*1.577*	*0.756*	*1.0*

*Norm., flow direction normal to bedding; Par., flow direction parallel to bedding; MS, medium swelling; HS, high swelling. Intrinsic permeabilities of Boom Clay were measured using a synthetic pore solution as a permeant. Values for bentonite are estimated from published data for pre-compacted Volclay Mx80 (Börgesson et al. 1996). Pre-test properties of bentonite are corrected for the probable effect of confining stress on the degree of saturation of the supplied material.

Table 2. *Summarized gas transport properties of Boom Clay and pre-compacted Volclay Mx80 bentonite*

Test number	Gas pumping rate, C (μl h^{-1})	Excess gas pressure, $(p_{gi} - p_{wo})$ (MPa)			Apparent matric suction, p_{co} (MPa)	Gas permeability, k_g (m$^2 \times 10^{20}$)		Confining stress, σ (MPa)	Back-pressure, p_{wo} (MPa)
		Breakthrough	Peak	Steady state		Flow in	Flow out		
T1S1	375	1.21	1.27	1.21	≈1.0	≈25.8	≈34.2	4.40	2.18
T2S1	375	1.88	1.89	1.73	≈1.0	7.9	6.5	4.40	2.18
T2S2	375	1.93	1.93	1.46	≈1.0	12.2	12.0	4.40	2.18
	180	—	—	1.38	≈1.0	7.0	8.3	4.40	2.18
	90	—	—	1.27	≈1.0	4.9	5.3	4.40	2.18
	45	—	—	1.25	≈1.0	2.6	4.7	4.40	2.18
	375	1.60	1.75	1.64	≈1.0	9.0	8.8	4.40	2.18
	180	—	—	1.42	≈1.0	6.4	5.1	4.40	2.18
	90	—	—	1.26	≈1.0	5.0	4.6	4.40	2.18
T3S3	375	3.57	3.59	3.15	2.1	5.4	5.1	7.40	2.21
	375	2.91	—	3.28	2.1	4.9	5.4	7.40	2.21
	375	3.04	3.37	3.28	2.1	4.9	4.6	7.40	2.21
	375	3.92	4.75	4.57	2.5	5.4	4.6	7.40	2.21
	$37\,500$	4.78	5.13	≈4.26	1.5	285	300	7.40	2.21
	3.75×10^6	—	—	—	—	≈28000	≈24000	7.40	2.21
T2S3	375	0.48	0.48	0.38	≈0.3	66.2	63.7	4.40	2.21
T3S1	375	1.01	1.02	0.82	0.5	9.7	9.4	4.40	2.21
	375	0.83	0.95	0.90	0.5	7.7	8.0	4.40	2.21
T3S2	375	0.38	1.08	1.06	≈0.2	3.6	3.7	4.40	2.21
	375	—	—	1.09	≈0.2	3.4	3.2	4.40	2.21
Mx80-1	375	14.92	15.17	13.74	11.8	2.9	2.7	16.00	1.01
Mx80-4A	375	15.19	15.30	14.22	11.0	1.8	1.7	16.00	1.01
	375	—	14.25	14.11	11.0	1.9	1.7	16.00	1.01
	180	—	—	13.64	11.0	1.1	0.9	16.00	1.01
	90	—	—	13.27	11.0	0.6	0.5	16.00	1.01
	45	—	—	12.91	11.0	0.4	0.3	16.00	1.01
Mx80-4B	375	15.37	15.78	15.47	13.2	2.5	2.2	18.00	1.01
Mx80-4C	375	17.01	17.01	16.01	13.0	1.9	2.0	20.00	1.01
Mx80-4D	375	17.91	17.95	16.79	14.1	2.1	1.7	22.00	1.01
Mx80-6*	375	7.04	7.04	6.92	≈4.0	2.2	1.6	8.00	1.01
	375	6.84	6.87	6.79	≈4.0	2.3	2.0	8.00	1.01
	180	—	—	6.71	≈4.0	1.1	0.9	8.00	1.01
	90	—	—	6.64	≈4.0	0.6	0.5	8.00	1.01
	45	—	—	6.56	≈4.0	0.3	0.2	8.00	1.01
Mx80-7	375	7.89	7.92	7.89	≈4.0	1.7	—	9.00	1.01
	375	8.01	8.02	7.98	≈4.0	1.7	1.4	9.00	1.01
	180	—	—	7.87	≈4.0	0.8	0.6	9.00	1.01
	90	—	—	7.80	≈4.0	0.4	0.3	9.00	1.01
	45	—	—	7.71	≈4.0	0.2	0.1	9.00	1.01

*Specimen contained an axial 'block fabrication joint'. Apparent matric suction is the excess gas pressure at the extrapolated asymptote of the shut-in transient. This extrapolation is rather uncertain in some experiments. The air entry value (AEV) is marginally lower than the excess gas pressure at breakthrough.

water interface cylinder and the tubing and porous disc were gas flushed. The initial gas pressure in the injection system was set using a regulator. After closing the valve to the regulator, the injection pump was set to constant flow rate mode. A very slow pumping rate of 375 μl h^{-1} was used in all tests to raise the upstream gas pressure to the point of initial gas breakthrough. The duration of the pressurization period varied from 6 to 20 days. Depending on the specific objectives of each test, the rate was then changed in stepwise manner to investigate the sensitivity of gas permeability to flow rate.

Data reduction

The principal stages of the controlled flow rate gas injection test can be defined as: (a) the pressure build-up stage; (b) the gas entry and pathway propagation stage; (c) the breakthrough stage; (d) the primary peak pressure response; (e) the spontaneous negative transient stage; (f) the steady-state stage; (g) the 'shut-in' response at zero flow rate (Horseman et al. 1996a). Additional stages may be added. By developing a differential equation for the rate of change of upstream gas pressure, a simple analytical data reduction model has been developed for the pre- and post-breakthrough stages, including the spontaneous negative transients, the steady states and the shut-in response (Ortiz et al. 1997). Although it is based on the rather restrictive assumption of constant gas permeability, the model provides a clear explanation of the post-breakthrough responses, a straightforward way of quantifying the gas transport variables, and an acceptable fit to certain subsets of the available data. At present, there is no model to describe the gas entry and pathway propagation stages, the breakthrough stage and the peak pressure response.

The governing differential equation of the testing apparatus is

$$V_{gi}\frac{dp_{gi}}{dt} + p_{gi}\frac{dV_{gi}}{dt} = B(p_{go}^2 - p_{gi}^2) \quad (11)$$

where p_{gi} and V_{gi} are the pressure and volume of gas in the upstream system and p_{go} is the gas pressure just inside the clay at the downstream end of the specimen. The transport variable B is given by

$$B = \frac{k_g A_s}{2\eta_g L_s} \quad (12)$$

where k_g is the effective gas permeability of the clay, η_g is the viscosity of helium, and A_s and L_s are the cross-sectional area and length of the specimen, respectively. Conventional two-phase flow theory would equate the effective gas permeability, k_g, with the product of intrinsic permeability, k, and relative permeability to gas, k_{rg}, according to equation (3). However, we shall demonstrate that this is problematic for clay-rich materials.

Downstream gas pressure, p_{go}, cannot be directly measured in the experiments. It should be related to the back-pressure of the water at the downstream end, p_{wo}, by the relationship

$$p_{go} = p_{wo} + p_{co}. \quad (13)$$

As this is similar to equation (9), we refer to p_{co} as the 'apparent' matric suction. For the purposes of data reduction this quantity is assumed to be constant. Analytical solution of equation (11) also demands that B be taken as constant and solutions for the various stages of an experimental history are given in the Appendix. Effective gas permeabilities at steady state are calculated using

$$k_g = \frac{2Q_{st}RT\eta_g L_s}{v_{mst}A_s[p_{gi}^2 - (p_{wo} + p_{co})^2]} \quad (14)$$

where Q_{st} (m^3 s^{-1}) and v_{mst} (m^3 mol^{-1}) are the volumetric flux and partial molar volume of the gas under STP conditions, R is the gas constant, and T is the temperature (Sen et al. 1996).

Permeabilities calculated in this way are somewhat model dependent and fairly sensitive to selected values of apparent matric suction. There are two independent measures of Q_{st}. The first is from the injection pumping rate and the second from the outward motion of the piston of the back-pressure pump. Permeabilities based on flow into a specimen are often a little higher than those from flow-out, possibly signifying some minor leakage of gas.

Results

All gas injection experiments show qualitatively similar responses with very little evidence of gas flow before the primary peak. Gas transport properties are summarized in Table 2. Boom Clay data have been reinterpreted and values vary a little from those previously quoted in progress reports (Volckaert et al. 1995; Ortiz et al. 1997). Typical behaviour is illustrated by Fig. 4, which shows the first stage of experimental history Mx80-4A on HS-batch bentonite. The excess pressure ($p_{gi} - p_{wo}$) of gas at the upstream face, referenced to back-pressure, is plotted against the elapsed time in seconds from the start of injection. Also shown are the STP volumetric flow rate into the upstream system (i.e. gas–water interface vessel, tubing and specimen) and the volumetric flow rate out of the specimen. Early in this test there is a slow build-up of upstream gas pressure with time. No measurable quantities of gas emerge from the downstream end of the specimen during this build-up. The response can be interpreted as a rise in injection gas pressure caused by slowly and isothermally compressing a fixed mass of gas in the upstream system. Gas breakthrough occurs at an excess gas pressure of 15.19 MPa,

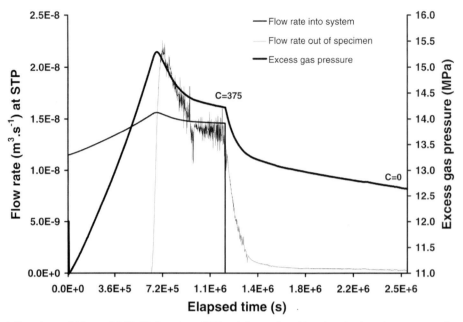

Fig. 4. Experimental history Mx80-4A Part 1 on HS-batch bentonite with a dry density of 1.669 Mg m^{-3} and a swelling pressure close to 15 MPa. Excess gas pressure and STP volumetric flow rates (into the testing system and out of the specimen) plotted against elapsed time. The peak pressure response is suggestive of crack propagation. C, gas pumping rate (µl h^{-1}).

which is fractionally larger than the effective stress on the clay

$$p_{gi} - p_{wo} > \sigma - p_{wo} = \sigma_{eff} = \Pi_{sw}. \quad (15)$$

For a high-swelling clay such as bentonite, the effective stress, σ_{eff}, can be directly equated with the swelling pressure, Π_{sw} (Horseman *et al.* 1996*b*). The peak response, at 15.30 MPa, is similar to that seen in hydrofracture experiments and is strongly suggestive of crack propagation (Murdoch 1993*a*). It is followed by a spontaneous and very well-defined negative transient which approaches an asymptote of 14.22 MPa. The gas injection pump was then stopped. Excess pressure follows a second negative pressure transient. The asymptote of this shut-in curve is the apparent matric suction, p_{co}, with a value around 11.0 MPa, which is well below the total stress.

Gas injection was reinstated at the standard pumping rate. Figure 5 shows no well-defined breakthrough event and gas flow through the specimen commences at an excess pressure substantially lower than that of the virgin clay. Gas pressure climbs to a rather inconspicuous secondary peak at an excess pressure of 14.25 MPa. This is 1.05 MPa lower than the primary peak. The post-peak negative transient is also rather poorly defined, with the steady-state asymptote around 14.11 MPa, which is a little lower than before. The explanation for the lack of a distinctive breakthrough event and the reduction in the peak excess pressure is that the gas pathways in the bentonite failed to reseal completely during the preceding shut-in stage. We suggest that this is due to the presence of bubbles or residual gas-filled voids left along the routes of gas migration, together with possible fabric damage.

Specimen Mx80-4 was then subjected to a descending history of pumping rates (180, 90, 45 and 0 µl h^{-1}). Each stage gives the negative transient response predicted by our data reduction model. As a rough rule of thumb, halving the gas flow rate at each stage leads to an excess pressure history with more or less equal decrements in the steady-state excess gas pressure. The gas flow law is very clearly nonlinear (i.e. non-Darcian) over this history, leading to variable gas permeability. The very substantial duration of the shut-in response is a direct consequence of this underlying nonlinearity. After 24 days of shut-in, the excess gas pressure falls to 11.60 MPa.

Conclusive evidence of gas penetration of the specimens was obtained by immersing previously gas-tested bentonite in a beaker of glycerol and

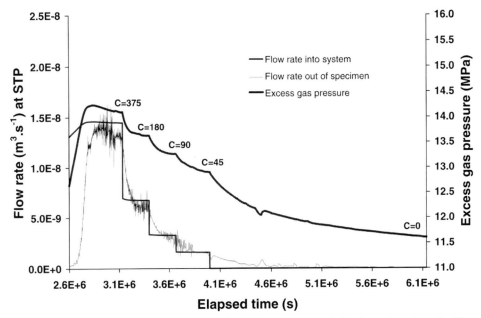

Fig. 5. Experimental history Mx80-4A Part 2 on HS-batch bentonite retested after the period of shut-in. The peak is substantially lower than for the virgin clay, suggesting that cracks formed in the first cycle were not fully closed at the onset of the second testing cycle. C, gas pumping rate (μl h^{-1}).

gently heating so as to expand and release helium trapped along the flow pathways. Streams of gas bubbles were observed (and photographed) emerging at numerous discrete points on all surfaces of the clay. As helium solubility in the clay pore water is extremely low, this shows that gas moves through the clay as a discrete phase, following multiple pathways. The interface between the sheath and the specimen appears to stay gas-tight, even at fairly elevated gas pressures. We believe that this interface is self-sealed by extensional strain (i.e. dilation) of the clay during gas entry. No emergent gas bubbles were observed in a control experiment on virgin bentonite.

Figure 6 shows experimental history T3S1 on Boom Clay and is a typical result for gas flow parallel to bedding in this overconsolidated mudrock. Early in this test there is the usual build-up of gas pressure with time. No measurable quantities of gas emerged from the downstream end of the sample during this period. Breakthrough is marked by a short period of gradually increasing downstream flow, followed by a dramatic rise in the flow rate. The pressure build-up curve terminates at a very well-defined peak, which is later than the breakthrough event. The excess gas pressure at breakthrough is 1.01 MPa.

The primary peak, at 1.02 MPa, is followed by a spontaneous negative transient with the pressure declining to its steady-state value. Close inspection reveals some evidence of a slow upward drift in pressure in the latter part of this transient. This has been observed many times and provides some insight into underlying system dynamics. At this point the injection pump was stopped. Excess gas pressure exhibits a second negative transient and declines towards its first shut-in value. The gas flux through the specimen is clearly intermittent in the later stages of this transient. Gas pumping was reinstated and pressure climbs to a broad secondary peak, somewhat lower than the primary peak, followed by another spontaneous negative transient. It is important to note that flow out of the specimen continued throughout the repressurization period. The new post-peak steady-state excess gas pressure is significantly higher than the original. The test was then completed with a final shut-in stage in which the pressure declines towards an equilibrium pressure close to its original value.

Gas permeabilities

Ignoring unrepresentative results, the effective gas permeability of Boom Clay normal to

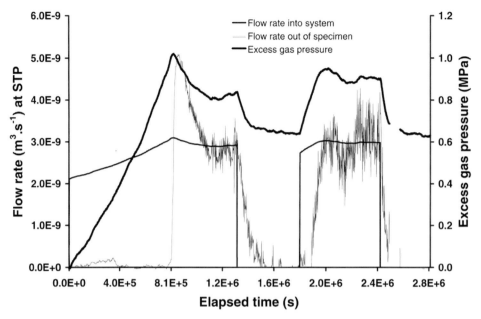

Fig. 6. Experimental history T3S1 on Boom Clay. This is a typical response for gas flow parallel to bedding in this overconsolidated clay. Breakthrough is marked by a short period of gradually increasing downstream flow, followed by a dramatic rise in flow rate. The gas flux is clearly intermittent in the later stages of the shut-in transient. The secondary peak is broader and somewhat lower than the primary peak.

bedding is found to be in the range $(5–12) \times 10^{-20}$ m^2, on the basis of flow-out data from tests at the standardized volumetric flow rate of 375 μl h^{-1} and a pre-test effective stress of 2.22 MPa. The equivalent range for flow parallel to bedding is $(3–9) \times 10^{-20}$ m^2. Gas permeabilities of both batches of bentonite are in the range $(2–3) \times 10^{-20}$ m^2, on the basis of tests at the standardized flow rate and effective stresses in the range 7–15 MPa. Gas permeabilities are clearly flow-rate dependent.

The gas permeability of Mx80-4 is plotted against net mean stress in Fig. 7. In unsaturated soil mechanics this quantity is the difference between the confining stress, σ, and the average internal gas pressure, taken as $(p_{gi} + p_{go})/2$. Net mean stress should be indicative of the tendency of the gas pathways to dilate or close with changes in their internal pressure. The permeability history starts at the gas breakthrough point, passes through the primary peak and evolves to the first steady state. The increase in gas permeability along the initial section of this line is indicative of decreasing resistance to flow, which results from the dilation of cracks. Maximum permeability occurs some time after the peak pressure. The shut-in transient is characterized by a very substantial fall in permeability. The zero flow-rate condition is never reached. It seems probable that the cross-sectional area of each of the gas-conducting pathways decreases throughout the period of shut-in. Gas flow continues throughout the repressurization period. The maximum permeability occurs at the steady-state condition and is almost identical to the steady-state value of the first cycle. The stepped history is marked by a substantial decline in permeability. It seems possible that, for repeated cycles of increasing and decreasing flow rate, the k_g v. net mean stress relationship settles into a unique, sigmoidal-shaped, hysteresis loop.

Gas permeabilities of Table 2 are of similar magnitude to intrinsic permeabilities of Table 1 based on tests using water as a permeant. Given the very low gas content of the samples after gas injection (a few per cent by volume only), these effective permeabilities to gas cannot be reconciled with the intrinsic permeabilities by invoking physically reasonable values for relative permeability to gas.

Significance of the peak

In experiments on Boom Clay, the primary peak occurs at gas pressures that are often significantly lower than the total isotropic stress applied to the specimen. The peak is also lower for flow parallel

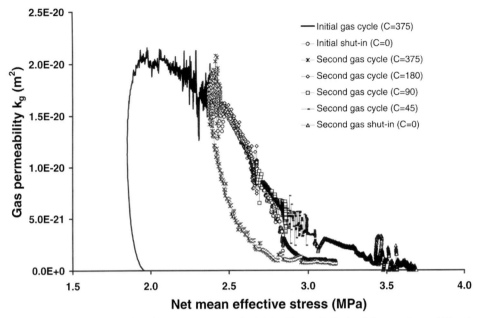

Fig. 7. Gas permeability plotted against net mean stress for Mx80-4A, on the basis of the experimental histories of Figs 4 and 5. The plot is indicative of pathway dilatancy after the breakthrough event and partial pathway closure during periods of declining gas pressure. The shut-in transients are each characterized by a substantial fall in gas permeability. C, gas pumping rate (μl h^{-1}).

to the bedding plane than for flow perpendicular to this plane. Although the overall shape of the excess pressure v. time curves is similar to that observed during hydrofracturing (Murdoch 1993a), the occurrence of hydraulic fractures at injection pressures less than total stress is unusual and current theory does not predict such a response. The primary peak is usually sharp and well defined. From the later stages of T3S3 (see Table 2), peak height is known to be sensitive to flow rate; this suggests that it must be at least partly determined by pressure gradients along the newly formed pathways. The secondary peak is more rounded and generally lower than the first; this feature shows that the shut-in periods were of insufficient duration for the pathways to entirely lose their gas by diffusion. We know that the full peak height can be regained by either increasing the total stress on the clay or flushing the pathways with water (Harrington & Horseman 1998). Re-establishment of the full peak is indicative of a capacity to self-seal.

Classic hydrofracture interpretation would suggest that the peaks in these curves represent the point at which the tensile strength of interparticle water films has been exceeded. On this basis, the tensile strength of the film (i.e. the adhesion force per unit area) would be the difference between the peak excess gas pressure and the excess pressure at shut-in. This quantity is direction dependent in Boom Clay and varies in the range 0.18–0.93 MPa for a confining pressure of 4.4 MPa. This is generally larger than the large-scale tensile strength of the material. The fact that shut-in pressure is less than total stress leads us to believe that the flow pathways are Mode 1 dilatant microcracks associated with extensile straining of the clay fabric when exposed to elevated gas pressures. The constant stress boundary condition distinguishes our tests from those performed using the pressure-plate apparatus. Under constant stress it is possible for a clay to exhibit extensile strain. Although the rupture mechanism is still under investigation, it seems to be the case that the high gas pressures lead to locally high pore-water pressures (Harrington & Horseman 1998). As the sum of effective stress and pore pressure then exceeds the boundary stress, there is an unbalanced internal tension imposed on the clay fabric. This internal tension is comparable with the film tension developed during clay shrinkage. Assuming linear elastic fracture mechanics (LEFM) to be applicable, it is possible that change in slope of the pressure–time curve just before the peak corresponds to the Griffith critical stress for the onset of stable crack propagation (Murdoch 1993b).

Shut-in pressure

The extrapolated asymptote of the shut-in curve (or apparent matric suction) represents some sort of lower threshold limit for gas flow in a clay. Setting aside diffusional transport, gas is never mobile at excess pressures less than this threshold. For Boom Clay, this pressure appears to be reasonably repeatable in tests on a single specimen, but varies from sample to sample and is generally lower for flow parallel to bedding than for flow in the other direction. For bentonite, the threshold pressure increases with swelling pressure and dry density. We suggest that the mechanical and surface tensional constraints on pathway stability make it impossible for gas to flow at excess pressures less than this threshold. Our data reduction model is based on the assumption that this lower threshold remains constant.

Dynamics

These experiments demonstrate that gas pathways in clay can be highly unstable. There is evidence of intermittent or 'burst-type' flow response in some of the pressure histories (see Fig. 6). Intermittent flow can be explained in terms of pathway propagation and collapse. When a gas pathway reaches the downstream end of the specimen, its internal pressure must drop sharply as it releases gas. As thin films of gas are inherently unstable, the drop in pressure will cause the pathway aperture to decrease. The bordering water films will coalesce in the more constricted regions. Continuous gas flow is then impossible. At this point, two possible mechanisms can be postulated. During continuous pumping, the blocking of pathways will lead to a rise in gas pressure in the upstream system. After some time, the pressure at the upstream end of the specimen will rise sufficiently for pathways to re-form. This can lead to the burst-type behaviour seen in some of our experiments. During a period of shut-ins it is not possible for the upstream gas pressure to rise. As intermittent flow does actually occur during a shut-in stage, gas pathways must be capable of slowly propagating through blockages under the action of a constant or gradually declining upstream gas pressure.

Desaturation

The amount of work that must be expended in countering the forces of water retention must vary locally in a clay; the variation depends largely on the proximity of the water molecules to the clay mineral surfaces. As we attempt to increase the gas saturation and decrease the volumetric water content, we must displace water from progressively thinner films, which is more strongly adsorbed to the clay minerals. Above the AEV, the matric suction of a clay increases very rapidly as the volumetric water content decreases (Delage & Graham 1995). This rapid rise in suction implies that exceptionally high gas pressures are necessary to produce large levels of gas saturation in compact clay-rich media subject to normal (i.e. positive) pore-water pressures. Under *in situ* conditions, it is improbable that gas pressure could exceed the total stress by any significant amount and, as a result, gas saturations more than a few per cent in normally pressured argillaceous media with high (>50%) clay contents are a physical impossibility.

Calculated water saturations, S_w, after gas injection into Boom Clay were generally $\geqslant 0.99$, with one exceptional value of 0.96. These are based on post-test gravimetric water contents and average particle specific gravity. Saturation estimates from routine water content measurements are generally imprecise, with second decimal place accuracies optimistic. Nevertheless, these calculated values clearly demonstrate that large volumes of gas can pass through clays and mudrocks without significant desaturation. This strongly contradicts the suggestion by Lineham (1989) that nearly all the water in a clay is displaced during gas injection. Although we cannot make the definitive statement that there was no displacement of water during gas movement, if it did occur then it was so small as to be immeasurable using our apparatus. There is, however, strong evidence from supplementary (unpublished) experiments on a consolidated bentonite paste that the overall gas content, in contrast to matrix gas saturation, of a clay does actually increase after breakthrough. If the gas content increases, but the water content of the matrix does not change significantly, then the only plausible explanation is that the overall volume of the specimen must increase.

Discussion

When the stress field is both uniform and isotropic, as in our experiments, gas pathways tend to propagate along the bedding planes in a naturally sedimented material such as the Boom Clay. Gas pressure at breakthrough is generally lower parallel to bedding than normal to bedding. Ease of pathway dilation provides a probable explanation. If the stress field is not isotropic, but the clay is homogeneous with no orientated fabric, then gas pathways will tend to propagate in a plane normal to the minor principal stress. Furthermore, there may be a tendency for pathways to follow a direction that coincides with the maximum stress gradient, with the gas moving down-gradient. As the effects of an anisotropic stress field and stress gradients have not been investigated experimentally, it is difficult to assess whether fabric anisotropy or stress would exercise the dominant control on the direction of gas migration in a homogeneous stratum of clay or mudrock.

As part of the EU MEGAS study (Volckaert et al. 1995; Ortiz et al. 1997), Volckaert and colleagues at SCK/CEN in Belgium conducted a number of gas injections into the Boom Clay in the HADES underground laboratory. The experiments involved slowly injecting gas into piezometers installed in boreholes at two locations within the laboratory, the first at the base of the main shaft and the second in the test drift. The piezometers had previously been used in long-term measurements of the pore-water pressures in this overconsolidated mudrock. Gas flow paths were from the injection filter to an adjacent, water-filled, monitoring filter within each cluster of piezometers. Gas seems to have moved largely within the zones of plastic deformation around the boreholes, suggesting that the destressing of the mudrock within these zones leads to a general lowering of gas entry pressure. Key observations from these field experiments were that: (a) breakthrough was comparatively rapid; (b) gas flow led to minimal desaturation; (c) gas flow occurred along one or more preferential pathways through the mudrock. These observations were broadly in line with the conclusions drawn from laboratory-scale experiments and from theoretical considerations. An interesting feature of these experiments, unexplained at the time, was the transmission of a pore-water pressure pulse to piezometers that were otherwise unaffected by the gas. We believe that this coupling between gas pressure and pore-water pressure is an important characteristic of the gas migration process in initially saturated clay-rich media.

Conclusions

It is our contention that initially water-saturated, compact clays and mudrocks do not behave as ordinary porous media, and that reliance on conventional two-phase flow theory is inadvisable when attempting to quantify gas-phase transport in these materials. Distinguishing features include the sub-microscopic dimensions of the interparticle spaces, the high specific surface of clay minerals, the strong physico-chemical interactions between water molecules and mineral surfaces, the deformable matrix, and the generally low tensile strength.

Gas does not occupy, or flow through, the intergranular pores of the clay matrix. In the absence of pressure-induced pathways, saturated clays and mudrocks are totally impermeable to gas. The measured gas permeability of such a clay or mudrock is therefore a dependent variable rather than a material property, as it depends on the number of pressure-induced pathways in the plane normal to flow, together with the width and aperture distributions of these features.

Passage of a gas phase into a pre-compacted and initially saturated bentonite is possible only if the gas pressure exceeds, by a small margin, the sum of the external equilibrium water pressure and the swelling pressure. Bentonite compacted to a high dry density will have a large swelling pressure and will display a commensurately large gas entry pressure. The extremely high gas entry pressure of bentonite in contrast to other clays can be anticipated from the studies of Croney & Coleman (1953) and Black (1962), who demonstrated that the AEV of a clay increases with increasing plasticity.

All injection experiments on virgin clay show a peak in gas pressure, followed by a spontaneous negative transient. The peak response is indicative of crack propagation, and this suggests that gas moves through an interconnected network of cracks that are formed by extensile rupture under high applied gas pressure. Crack dilation and closure provide an explanation of the nonlinearity of the flow law revealed by multi-stage testing. When gas injection stops, the pressure in the pathways spontaneously declines and asymptotically approaches some finite value. No gas flow has ever been detected at an excess gas pressure lower than this critical lower limit. We suggest that mechanical and surface tensional constraints on pathway stability make it impossible for gas to flow at an excess pressure that is lower than this threshold.

When gas flow is re-established, the threshold pressure for gas breakthrough is found to be significantly lower than in the virgin clay. This is indicative of the presence of residual gas-filled voids left along the flow pathways, and of possible fabric damage. It is known that the high breakthrough pressure and associated primary peak response of a virgin clay specimen can be regained either by increasing the total stress acting on the clay or by flushing the flow pathways with water.

On the basis of our brief examination of the principle of axis-translation, we suggest some degree of commonality between the mechanisms of gas entry and penetration of an initially water-saturated clay or mudrock subject to compressive boundary stresses and positive pore-water pressures (i.e. conditions at some depth below ground surface) and those that lead to the development of gas-filled cracks in clays subject to desiccation, shrinkage and negative pore-water pressures. The experiments support the hypothesis that transport pathways are Mode 1 (dilatant) microcracks associated with the extensile straining of the clay fabric. It seems to be the case that the high gas pressures in our experiments cause locally high pore-water pressures. As the sum of effective stress and pore pressure then exceeds boundary stress, an unbalanced internal tension is imposed on the clay fabric.

Studies on Boom Clay were funded as a component of the MEGAS (EC PEGASUS) Project (Contract F12W-CT91-0076) of the 3rd Framework R&D Programme of the European Union. Research on bentonite was funded by the Swedish Nuclear Fuel and Waste Management Co. (SKB), Stockholm. Additional work on gas migration in landfill-liner clays is supported by the EPSRC/NERC Waste Management and Pollution Programme. In particular, we thank R. Marsden of the Royal School of Mines, Imperial College, London, who supervised the first author's external PhD activities on fluid flow in mudrocks. We also wish to thank our sponsors and our research partners for the continuing and sometimes lively debate on matters of technical importance to waste management and disposal. This paper is published with the permission of the Director, British Geological Survey (NERC).

References

AZIZ, K. & SETTARI, A. 1979. *Petroleum Reservoir Simulation*. Applied Science, London.

BLACK, W. P. M. 1962. A method of estimating the California Bearing Ratio of cohesive soils from plasticity data. *Géotechnique*, **12**, 271–282.

BÖRGESSON, L., KARNLAND, O. & JOHANNESSON, L.-E. 1996. Modelling of the physical behaviour of clay barriers close to water saturation. *Engineering Geology*, **41**, 127–144.

CRONEY, D. & COLEMAN, J. D. 1953. Soil moisture suction properties and their bearing on moisture distribution in soils. *Proceedings of the 3rd International Conference on Soil Mechanics and Foundation Engineering*, 13–18.

DELAGE, P. & GRAHAM, J. 1995. Mechanical behaviour of unsaturated soils: understanding the behaviour of unsaturated soils requires reliable conceptual models. *In*: ALONSO, E. E. & DELAGE, P. (eds) *Unsaturated Soils. Proceedings of the 1st International Conference on Unsaturated Soils, Paris*. Balkema, Rotterdam, 1223–1256.

DE MARSILY, G. 1986. *Quantitative Hydrogeology for Engineers*. Academic Press, New York.

DUPPENBECKER, S. J., DOHMEN, L. & WELTE, D. H. 1991. Numerical modelling of petroleum expulsion in two areas of the Lower Saxony Basin, Northern Germany. *In*: ENGLAND, W. A. & FLEET, A. J. (eds) *Petroleum Migration*. Geological Society, London, Special Publications, **59**, 47–64.

FREDLUND, D. G. & RAHARDJO, H. 1993. *Soil Mechanics for Unsaturated Soils*. Wiley, New York.

HARRINGTON, J. F. & HORSEMAN, S. T. 1998. *Diagnostic experiments on gas migration in clay*. Paper presented at the EU PROGRESS Project Workshop, Naantali, Finland, 25–27 May. British Geological Survey Technical Report **WE 98/28**.

HEDBERG, H. D. 1974. Relation of methane generation to undercompacted shales, shale diapirs and mud volcanoes. *Bulletin, American Association of Petroleum Geologists*, **58**, 661–673.

HORSEMAN, S. T., HARRINGTON, J. F. & SELLIN, P. 1996a. Gas migration in Mx80 buffer bentonite. *Symposium on the Scientific Basis for Nuclear Waste Management XX, Fall Meeting, 2–6 December, Boston*. Materials Research Society, Pittsburgh, PA, 1003–1010.

——, HIGGO, J. J. W., ALEXANDER, J. & HARRINGTON, J. F. 1996b. *Water, gas and solute movement in argillaceous media*. Prepared by British Geological Survey for NEA SEDE Working Group on Measurement and Physical Understanding of Groundwater Flow Through Argillaceous Media. OECD Nuclear Energy Agency, Report **CC-96/1**.

JOHANNESSON, L.-E., BÖRGESSON, L. & SANDÉN, T. 1995. *Compaction of bentonite blocks: development of technique for industrial production of blocks which are manageable by man*. Swedish Nuclear Fuel and Waste Management Co. (SKB), Technical Report **95-19**.

LINEHAM, T. R. 1989. *A laboratory study of gas transport through intact clay samples*. Nirex Safety Studies Report **NSS-R155**.

MARSHALL, T. J. & HOLMES, J. W. 1979. *Soil Physics*. Cambridge University Press, Cambridge.

MURDOCH, L. C. 1993a. Hydraulic fracturing of soil during laboratory experiments. Part 1. Methods and observations. *Géotechnique*, **43**, 255–265.

——1993b. Hydraulic fracturing of soil during laboratory experiments. Part 3. Theoretical analysis. *Géotechnique*, **43**, 277–287.

NEWMAN, A. C. D. 1987. The interaction of water with clay mineral surfaces. *In*: NEWMAN, A. D. C. (ed.) *Chemistry of Clays and Clay Minerals*. Mineralogy Society Monograph, **6**, 237–274.

ORTIZ, L., VOLCKAERT, G., DE CANNIERE, P. *et al.* 1997. MEGAS — modelling and experiments on gas migration in repository rocks: Final Report — Phase 2. European Commission, Nuclear Science and Technology, **EUR 17453 EN**.

SEN, M. A., HORSEMAN, S. T. & HARRINGTON, J. F. 1996. Further studies of the movement of gas and water in an overconsolidated clay. EC MEGAS

Project — Phase 2. British Geological Survey Technical Report **WE/96/8**.
Tissot, B. & Pelet, R. 1971. Nouvelles données sur les mécanismes de genèse et de migration du pétrole: simulation mathématique et application à la prospection. In: *Proceedings of the 8th World Petroleum Congress, Moscow*, 35–46.
Volckaert, G., Ortiz, L., De Canniere, M. *et al.* 1995. MEGAS — Modelling and experiments on gas migration in repository rocks: Final Report — Phase 1. European Commission, Nuclear Science and Technology, **EUR 16235 EN**.

Appendix: data reduction model

Assuming ideal gas behaviour, the post-breakthrough laminar flux of a compressible gas phase passing through cross-sectional area A_s of a laboratory specimen of length L_s is given by

$$Q_{st} = \frac{v_{mst} k_g A_s}{2RT\eta_g L_s}(p_{gi}^2 - p_{go}^2) \quad (A1)$$

where Q_{st} is the volumetric flux (m³ s⁻¹) under standard temperature and pressure (STP) conditions, v_{mst} (=0.0224 m³ mol⁻¹) is the partial molar volume of the gas at STP, k_g (m²) is the effective gas permeability of the specimen, R (8.314 J mol⁻¹ K⁻¹) is the gas constant, T (K) is the temperature, η_g (Pa s) is the viscosity of the gas, p_{gi} (Pa) is the pressure of the gas just inside the specimen at the upstream end, and p_{go} is the pressure of the gas just inside the specimen at the downstream end (Ortiz *et al.* 1997). The downstream gas pressure is assumed to be related to the back-pressure of the water at the downstream end, p_{wo}, by the relationship

$$p_{go} = p_{wo} + p_{co} \quad (A2)$$

where p_{co} is the apparent value of the matric suction. The governing differential equation for the axial flow apparatus is given by

$$V_{gi}\frac{dp_{gi}}{dt} + p_{gi}\frac{dV_{gi}}{dt} = B(p_{go}^2 - p_{gi}^2) \quad (A3)$$

where p_{gi} and V_{gi} are the pressure and volume of gas in the upstream system and B is a transport variable given by

$$B = \frac{k_g A_s}{2\eta_g L_s}. \quad (A4)$$

Analytical solution of the differential equation demands that B and p_{co} be taken as constants (Sen *et al.* 1996; Ortiz *et al.* 1997). The upstream pressure build-up at a constant volumetric pumping rate (C is constant), before gas breakthrough has occurred, can be quantified by taking $B = 0$. The solution is

$$p_{gi} = p_{gi0}\left(\frac{V_{gi0}}{V_{gi0} - Ct}\right) \quad (A5)$$

where V_{gi0} and p_{gi0} are the initial volume and pressure of gas in the gas injection (upstream) system, C (m³ s⁻¹) is the volumetric pumping rate and t (s) is the elapsed time from the start of pumping. The upstream pressure build-up therefore depends on the initial volume and pressure of the gas in the upstream system and on the volumetric pumping rate.

If the syringe pump is operating at a constant rate (i.e. C is constant) then

$$V_{gi} = V_{gi0} - Ct \quad \text{and} \quad \frac{dV_{gi}}{dt} = -C \quad (A6)$$

where t is the elapsed time after the change in pumping rate. This leads to

$$(V_{gi0} - Ct)\frac{dp_{gi}}{dt} = B(p_{go}^2 - p_{gi}^2) + Cp_{gi}. \quad (A7)$$

The solution is given by

$$p_{gi} = \frac{\left[E(p_{gi0} - D) - D(p_{gi0} - E)\left(\frac{V_{gi0}}{V_{gi0} - Ct}\right)^G\right]}{\left[(p_{gi0} - D) - (p_{gi0} - E)\left(\frac{V_{gi0}}{V_{gi0} - Ct}\right)^G\right]} \quad (A8)$$

where

$$D = \frac{C}{2B} + \sqrt{\left[\left(\frac{C}{2B}\right)^2 + p_{go}^2\right]},$$
$$E = \frac{C}{2B} - \sqrt{\left[\left(\frac{C}{2B}\right)^2 + p_{go}^2\right]} \quad (A9)$$

and

$$G = \frac{B(D - E)}{C}. \quad (A10)$$

Relationship (A8) therefore describes the evolution, with time, of the upstream gas pressure from the initial condition $p_{gi} = p_{gi0}$ at $t = 0$ (some time after breakthrough has occurred), with the injection syringe pump running at a constant and finite rate (i.e. C is constant).

For constant volumetric pumping rate (i.e. C is constant), the upstream gas pressure p_{gi} evolves to a steady-state value characterized by

$$\frac{dp_{gi}}{dt} = 0. \quad (A11)$$

The steady-state upstream gas pressure is then given by

$$p_{gi} = \frac{C}{2B} + \sqrt{\left[\left(\frac{C}{2B}\right)^2 + p_{go}^2\right]}. \quad (A12)$$

For a fixed value of B, the upstream gas pressure at steady state should therefore depend on the volumetric pumping rate, C, and on the downstream gas pressure p_{go}, and should be independent of test path followed to achieve the steady-state condition. According to the model, upstream gas pressure at steady state should be a repeatable quantity. If, after breakthrough, the syringe pump is stopped ($C = 0$), the upstream gas pressure will decay with time. Solution of the differential equation gives

$$p_{gi} = p_{go} \frac{\left[(p_{gi0} + p_{go})\exp(Ht) + (p_{gi0} - p_{go})\right]}{\left[(p_{gi0} + p_{go})\exp(Ht) - (p_{gi0} - p_{go})\right]} \quad (A13)$$

where t is the elapsed time from stopping the pump and

$$H = \frac{2Bp_{go}}{V_{gi0}}. \quad (A14)$$

The upstream gas pressure, p_{gi}, therefore asymptotically approaches the downstream gas pressure, p_{go}, and after infinite elapsed time

$$p_{gi}(t = \infty) = p_{go} = (p_{wo} + p_{co}) = \text{constant}. \quad (A15)$$

As p_{wo} remains constant in our experiments, determination of the upstream gas pressure at the asymptote provides a method of quantifying p_{co}. The decay time of the transient is determined by H, and depends on the initial volume and pressure of the gas in the injection (upstream) system and on the gas permeability of the clay. Large gas volumes in the upstream system lead to very lengthy experimental transients.

Top-seal leakage through faults and fractures: the role of mudrock properties

GARY M. INGRAM[1] & JANOS L. URAI[2]

[1]*Shell Development (Australia) Pty Ltd, Shell House, QV1 Building, 250 St George's Terrace, Perth, WA 6000, Australia (e-mail: Gary.Ingram@shell.co.uk)*
[2]*RWTH Aachen, Lochnerstrasse 4-20, D-52056 Aachen, Germany (e-mail: j.urai@ged.rwth-aachen.de)*

Abstract: Mudrocks are effective top seals for hydrocarbon accumulations because they possess very low permeabilities, high capillary entry pressures, and are often laterally continuous basin-wide. For leakage through the seal to take place, an additional mechanism must provide enhanced permeability in mudrocks. Tectonically induced, dilatant faulting and fracturing in brittle rocks is such a mechanism. The effectiveness of mudrocks as seals may be compromised by a number of other factors, such as tectonic fault displacements in excess of the seal thickness; tensile fracturing under extreme fluid pressure conditions; and leakage via a network of juxtaposed thin leaky beds across sub-seismic faults within the seal. Before hydrocarbon trap integrity analysis, seismic interpretations should honour the fundamental geometry of the trap as closely as possible because top seals can be reliably appraised only when basic geometries are accurately determined. Fluid pressure is a proven risk in many exploration provinces, in terms of mudrock top-seal leakage via opening mode fractures. This natural hydrofracturing can take place if buoyancy pressures, combined with fluid overpressures, exceed the minimum *in situ* horizontal stress, plus the tensile strength of the seal, an occurrence that leads to mechanical failure of the seal rock. The rheology of a seal is another important factor as it determines the failure mode, i.e. whether the rocks are ductile and remain sealing after deformation or whether they deform in a brittle manner to create permeable leak paths. Several techniques have been developed to predict rock rheology; these rely on uplift, sonic velocity and clay content data. Burial curves can be used to determine overconsolidation of mudrocks. Sonic velocity data can be used to estimate unconfined compressive strength in mudrocks. The most direct method utilizes relationships between mudrock friction angle, swelling clay content and mudrock surface area to determine ductility. The effect on trap integrity of sub-seismic faults is also quantifiable because advances in structural geological technology now permit predictions of the numbers of sub-seismic faults in a trap. This information, combined with detailed top-seal stratigraphic data, provides the power to screen traps for the risk of top-seal leakage via sub-seismic fault juxtapositions within the seal. A simple strategy incorporating these factors and linking them to other structural and stratigraphic information can contribute to reducing uncertainty about top-seal leakage caused by tectonic deformation.

Many effective top seals of hydrocarbon accumulations are mudrocks. They typically have extremely low permeabilities and correspondingly high capillary entry pressures. Ductility is a key property to ensure effective mudrock seal behaviour when a trap is influenced by tectonic deformation (Downey 1984; Grunau 1987).

The limits of capillary sealing

Lithology has an important control on sealing capacity of top seals. Although there is a trend from high capillary seal capacity in fine-grained, clay-rich rocks to low seal capacity in coarse-grained, clay-poor rocks, under conditions of active charge and leakage, the maximum hydrocarbon column height of a trap will be regulated by capillary displacement pressure only where top seals are relatively permeable. Permeability in a capillary seal is necessary to allow rapid 'bleeding-off' of hydrocarbons through the seal pore space when the capillary entry pressure is exceeded, so that the hydrocarbon pressure is rapidly brought below the capillary seal capacity again and leakage ceases.

According to capillary sealing theory (Schowalter 1981; Watts 1987), tight mudrocks can hold very large hydrocarbon columns. However, pore diameters in such mudrocks have the same dimensions as larger oil molecules (nanometres) and the interfacial physics controlling capillary effects may indeed break down or become irrelevant. Thus equations relating to capillary forces in oil–water systems may not hold any more. In tight mudrocks such as these, it is likely that leakage of oil cannot occur by Darcy flow through matrix porosity (Hall et al. 1986) and therefore leakage through fracture networks is the most likely leak mechanism. Diffusion of light hydrocarbons through the stationary pore fluid in mudrocks may represent an additional leak mechanism for gas (Schlomer & Krooss 1997). To demonstrate in a simple way the effectiveness of fractures as conduits for fluid flow, Table 1 shows some example calculations for leakage through fractures with two different aperture sizes, simplified to Pouseille flow.

At extremely low permeabilities (in the nano-Darcy range), such as those found in pure mudrocks (e.g. van Oort et al. 1996), the capillary entry pressure will commonly seal hydrocarbon columns much larger than the maximum possible closure and therefore the only seal risk comes from the formation of dilatant hydrofractures.

Leakage through two different types of dilatant fractures, extensional fractures and tectonic shear fractures, will now be considered. These types of fractures can cause the formation of open, permeable pathways within sealing formations, most markedly in strong, cemented formations.

Extension fractures

If the condition $P_f = \sigma_3 + T$ is satisfied, extension fractures (Fig. 1) will form in a top seal and allow highly pressurized fluids to escape (P_f is the fluid pressure, σ_3 the minimum principal (total) stress and T the tensile strength of the seal). This situation will occur in highly overpressured systems, where interaction of the *in situ* stress field and local variations in fluid pressure (e.g. pressure differences between reservoir and seal) determine retention (e.g. Gaarenstroom et al. 1992). The key factor to consider here is retention capacity ($\sigma_3 - P_f$). Prospects with a low retention capacity have a high chance of seal failure.

For extension fractures formed by hydrocarbon intrusion into the seal, the leak criterion is $P_{HC} = \sigma_3 + T$ (P_{HC} is the hydrocarbon pressure at the top-seal–reservoir interface). It is worth noting that for more permeable lithologies, one can calculate a maximum depth, Z_{max}, below which capillary entry of the hydrocarbon phase into the seal is easier than formation of extension fractures (Du Rouchet 1981; Mandl & Harkness 1987):

$$Z_{max} = 0.04 \left\{ \frac{1 - 2v}{1 - \lambda} \times \frac{P_{dhc}}{K'[(1 - v)(-v)]} \right\} \quad (1)$$

where P_{dHc} is the hydrocarbon–water displacement pressure of the seal, λ is overpressure ratio, i.e. P_f/σ_v (where σ_v is the vertical stress), K' is the stress ratio (σ_h/σ_v) and v is Poisson's ratio. The formation of extension fractures becomes increasingly more suppressed in rocks with relatively low P_{dHc}, and at high K' and λ.

Table 1. *Leak rates through a fracture*

		Fracture aperture, a (m)	
Variable	Unit	1×10^{-5}	1×10^{-6}
Oil viscosity, η_o	Pa s	0.01	0.01
Gas viscosity, η_g	Pa s	3×10^{-5}	3×10^{-5}
Fracture width, w	m	10	10
Fracture length, Δl	m	50	50
Pressure difference, Δp	MPa	1.0	1.0
Leak rates: Q	m^3 s^{-1}	2×10^{-8}	2.0×10^{-11}
Q_{oil}	m^3 Ma^{-1}	6.3×10^5	6.3×10^2
Q_{gas}	m^3 Ma^{-1}	6.3×10^8	2.1×10^5
Q_{gas}	ft^3 Ma^{-1}	2.2×10^{10}	7.4×10^6

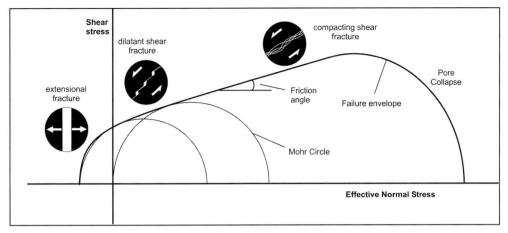

Fig. 1. Mode of fracture formation, depicted on a Mohr diagram. Extensional fractures form in rocks when the least normal stress becomes tensile enough to overcome the minimum *in situ* stress and rock cohesion. Hybrid, dilatant, shear fractures occur in rocks during deformation in low confining stress regimes and in rocks with a high friction angle (relatively strong rocks). Compacting shear fractures occur during deformation of rocks at high confining pressures (under deep burial) or during deformation of weak, ductile, rocks (low friction angle). Only compacting shear fractures will remain closed and sealing after deformation.

Shear fractures

In normally pressured systems, dilatant shear fractures may form in top seals, and thus it is not always necessary to have extensional fractures forming in response to high fluid pressure, for leakage to take place. The formation of extensional shear fractures is dependent on the size and position of the Mohr stress circle (which describes the *in situ* stress) relative to the failure envelope of the seal (Fig. 1). Limiting conditions can be approximated using a modified Mohr–Coulomb failure criterion, rewritten in terms of effective principle stresses at failure:

$$\sigma'_1 = 2CF^{0.5} + F\sigma'_3 \qquad (2)$$

where C is cohesion, $F = [(1 + \mu^2)0.5 + \mu]^2$ with μ the coefficient of friction, and σ'_1 is the largest principle effective stress. By defining a tensile strength T ($T < 0.5C$), conditions for the simultaneous occurrence of extensional and shear fractures (i.e. a leaky seal) can be expressed as a function of overpressure ratio v. depth (Sibson 1981; Behrmann 1991) to allow distinction between the fields of shear and extensional fracture. All other factors remaining constant, the risk of dilatancy can be expected to increase with decreasing effective pressure (i.e. the Mohr circle will move towards the left and the extensional field).

Brittle and ductile behaviour

In this paper we define (somewhat unconventionally) a *ductile* mudrock as one that can deform without dilatancy and the associated creation of fracture permeability. A *brittle* mudrock is defined as one that dilates during deformation and allows fracture permeability to develop (Urai & Wong 1994; Urai 1995). Thus a brittle mudrock is anomalously strong compared with normally consolidated rocks at the same depth. The key factor determining the onset of dilatancy is the ratio of compressive strength and mean effective stress:

$$\frac{(\sigma_1 - \sigma_3)}{(\sigma'_1 + \sigma'_2 + \sigma'_3)} \qquad (3)$$

(see Wood (1990) for a full discussion).

Under which geological conditions does a mudrock dilate during deformation?

The main micro-scale controls on dilatancy during shear failure in mudrocks can be illustrated by considering the processes in a small rock element, as it passes through the stress–strain curve (Fig. 2). Strain begins to localize in the sample at peak differential stress and this effect leads to the formation of undulating shear zones (shear fractures). Sliding on these surfaces

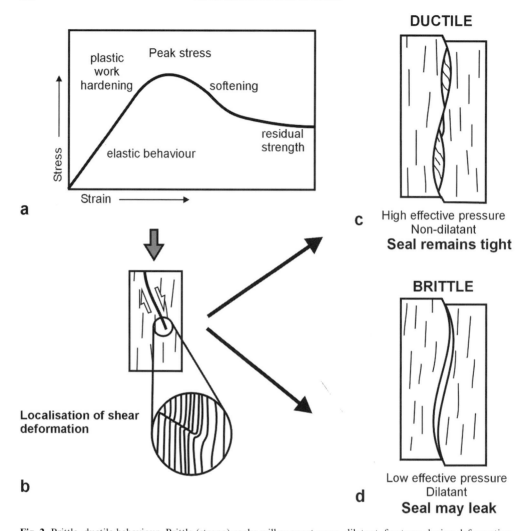

Fig. 2. Brittle–ductile behaviour. Brittle (strong) rocks will support open, dilatant, fractures during deformation, thus representing a leak risk. (**a**) An idealized stress–strain curve for shales deformed in triaxial compression. (**b**) Macroscopic failure modes of sandstones and shales. (**c**) & (**d**) Schematic illustration of the micro-scale processes leading to dilatant or non-dilatant deformation in poorly cemented shales.

accommodates more deformation but, because of their curvature, sliding can be accommodated only by either dilatancy or by formation of new shear zones. Fundamentally, the tendency to dilate will be a function of (1) mechanical properties of the rock, (2) effective pressure and (3) shear-zone geometry. At a given effective pressure a stronger (overconsolidated or cemented) rock is more likely to dilate than a weaker one.

Evaluation strategies of seal embrittlement

To recap, the key to predicting embrittlement of mudrock top seals is the prediction of the conditions where dilatant shear failure occurs. Three methods may be used for estimating seal embrittlement in mudrocks: (1) overconsolidation ratio; (2) mudrock brittleness index (BRI); (3) determination of friction angle from surface area.

Estimation of ductility from overconsolidation ratio

The overconsolidation ratio (OCR) is a relatively simple approach to rock strength determination and is useful in situations where few data are present other than interpretations of burial history from stratigraphic relations. The OCR is essentially a measure of uplift and may be described by the following equation:

$$OCR = \frac{P_{eff}max}{P_{eff}now} \quad (4)$$

where OCR is the overconsolidation ratio, $P_{eff}max$ is the maximum past effective pressure and $P_{eff}now$ is the present-day effective pressure.

Rock mechanical tests indicated that the critical OCR needed for dilatancy in uncemented mudrocks is significantly larger than unity. The results of these tests indicated that normally compacted, uncemented mudrocks can be expected to be ductile over the whole depth range, and only very strong uplift will lead to embrittlement. However, some cemented mudrocks were found to dilate at OCR of unity. Cementation can be separated into two types: localized and pervasive. Shallow cementation (phosphate or calcite nodules; Smith 1987) is usually localized and strongly influenced by depositional environment: it is unusual to find the whole mudrock body cemented to such an extent that fractures may extend through the whole seal. Therefore, the risk of embrittlement by this process is in general low. On the other hand, pervasive cementation in mudrocks is a mainly depth-dependent process. With increasing burial (also depending on environment), mobility of silica and calcite is increased, phyllosilicates start to recrystallize and the mudrocks are increasingly lithified. This process is accompanied by diagenetic changes in the clay minerals, such as the smectite–illite transition. Cementation increases the rock's strength in comparison with its uncemented equivalent at the same depth, and it is expected that these stronger mudrocks will dilate when deformed and represent a leak risk.

Estimation of mudrock brittleness from the rock's unconfined compressive strength (UCS)

This method uses an engineering approach to equation (3) and defines the brittleness index BRI = UCS/UCS_{NC}, where UCS_{NC} is the unconfined compressive strength of a normally consolidated rock in non-overpressured domains. UCS can be measured directly on plug samples or estimated from logs or cuttings on the basis of empirical correlations (Urai 1995) (Fig. 3), expressed by the following equations:

$$\log UCS = -6.36 + 2.45 \log(0.86V_p - 1172) \quad (5)$$

$$\log UCS = -6.36 + 2.45 \log V_s \quad (6)$$

with UCS in MPa, and V_p and V_s in m s^{-1}.

UCS_{NC} is determined from empirical soil mechanics correlations and is estimated (for clay-rich soils) by the equation: $UCS_{NC} = 0.5\sigma'$, where σ' is the *in situ* effective pressure corresponding to normal consolidation at the depth of interest (Horseman *et al.* 1986). Admittedly, this is an approximation only to be used if no further data are available.

For BRI > 2 the risk of embrittlement increases with increasing BRI (Fig. 4). However, the risk of development of an open fracture network cutting the whole seal depends on more factors than local seal strength and therefore the BRI criterion is likely to be conservative, so that a seal classified as brittle may still retain hydrocarbons. The power of this approach is that it can be applied on a regional scale, by constructing a BRI map of a given formation from a collection of sonic logs.

Prediction of mudrock friction angle from surface area

If the friction angle of a mudrock top seal is known, a constraint can be placed on the minimum stress at which the seal fails. Such information can be used for ranking prospects in order of risk of dilatant fracturing. Friction angle can be predicted from the rock's surface area (as measured by the dielectric constant (DCM) method; Fig. 5), which can be used as a measure of the total fraction of swelling clays (Steiger & Lueng 1988; Olgaard *et al.* 1999). Surface area measurement can be done using cuttings from existing wells, and the friction angle may be estimated from the surface area using the following equation (Urai *et al.* 1997):

$$\log \phi = \log 35 - S(\log 35/1266) \quad (7)$$

where ϕ is the angle of internal friction.

Between surface area values of 100 and 400 m^2 g^{-1} (most downhole shales correspond to this range of DCM-derived values) there is scatter in the data, representing an error of ±15–20%. This scatter was caused by variation in mechanical properties of specimens when loaded in different directions.

If only surface area information from cuttings is available, high surface area (e.g. >300 m^2 g^{-1})

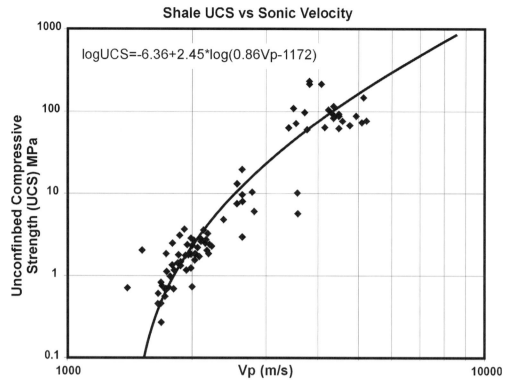

Fig. 3. Unconfined compressive strength (UCS) and brittleness. UCS v. V_p from a combined data set. This crossplot indicates that UCS is related to sonic velocity in a predictable manner. Note that at low V_p, UCS is expected to decrease more strongly, following an asymptotic trend of V_p to water.

can be taken to indicate the absence of cementation and thus a ductile mudrock. If both surface area and sonic velocity data are available, one can calculate the 'true' brittleness index at the *in situ* conditions of a given seal.

Layered mudrock–silt top seals

The top seals of many hydrocarbon accumulations consist of mudrock layers alternating with thin, permeable siltstones and sandstones. Such seals may leak if faults are sufficiently numerous, and their throws great enough, to create a connected network of juxtaposed leaky beds (Fig. 6), in the absence of fault seals.

Fault-assisted top-seal leakage

The risk of leakage, through a fault-connected network of leaky beds, can be quantified from the number of relatively thick mudrock beds in the seal and the statistics of the fault population in the trap area, derived from 3D seismic data (Ingram *et al.* 1997).

To model fault-assisted top-seal leakage, a basic configuration of a number of identical mudrock layers of similar thickness, in which faults are randomly dispersed, was considered. The mudrock layers are defined in this context as mudrock strata that are separated by very thin, laterally continuous, beds of relatively leaky lithologies, such as siltstones and sandstones. In this model, fault throws may juxtapose adjacent leaky beds, but the faults may not seal (there is no clay smear or cataclasis), or act as conduits to flow. The probability of leakage can be found by considering the permutations of fault positions relative to the mudrock layers in the seal. Figure 7 shows the analytical solution for probability of top-seal leakage.

The risk of fault-assisted top-seal leakage may be estimated by calculating the throw–frequency distribution for the rock volume around the seal from length v. cumulative frequency statistics. To do this, fault lengths must be converted to throws and the fractal dimension of fault populations should be adjusted relative to sample dimensions (the fractal dimension of the popu-

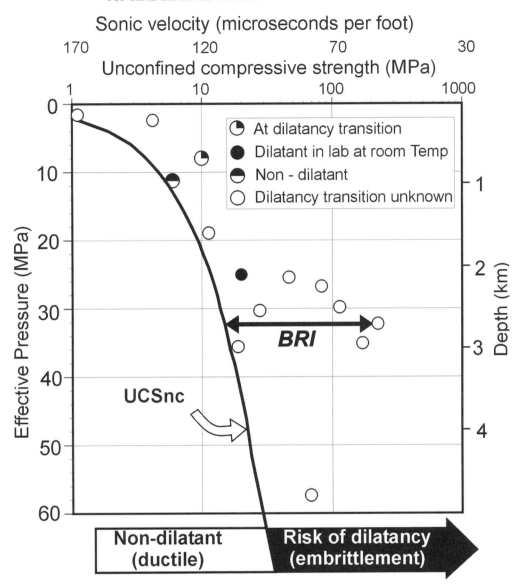

Fig. 4. Brittleness index. A plot showing the use of Brittleness index in the evaluation of massive shale top seals. Brittleness index, BRI = UCS/UCS_{NC}, where UCS_{NC} is the UCS of normally consolidated shales. Above BRI = 2, the likelihood of dilatant behaviour increases.

lation is related to the geometric dimensions of the sampling domain, i.e. areas have a fractal dimension of two, and volumes have a dimension of three), assuming that the map samples the largest fault close to its true maximum throw (Gauthier & Lake 1993).

By comparing the trap volume fault population with the fault population required to provide sufficiently numerous sub-seismic faults for leakage in the top seal, the seal risk may be determined (Ingram et al. 1997). As a general rule, this model predicts that leak probabilities of >50% and >90% are reached when the number of faults of appropriate size for juxtaposition (i.e. they have maximum throws that are equal to, or slightly greater than, the thickness of the component mudrock layers) is four and six times the number of mudrock layers in the seal.

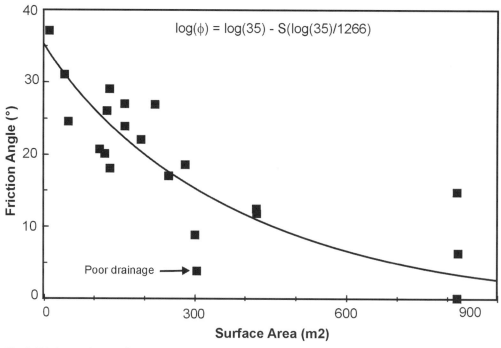

Fig. 5. Friction angle v. surface area relation. A relationship between surface area of mudrocks and their friction angle is indicated by the above cross plot. All the samples shown in the plot were deformed under controlled laboratory conditions and surface area was measured via the dielectric constant method. A regression has been derived which permits the determination of rock strength from surface area data alone.

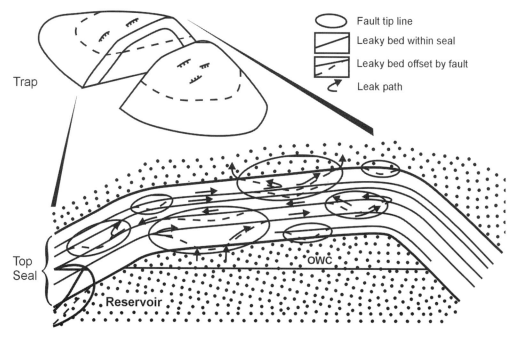

Fig. 6. Fault-linked leak path. Small, sub-seismic, faults may link up leaky strata in a top seal, thus forming a tortuous, but effective, leak pathway over geological time. The mechanism requires that the sub-seismic faults do not form clay smears or cataclastic seals against the leaky strata.

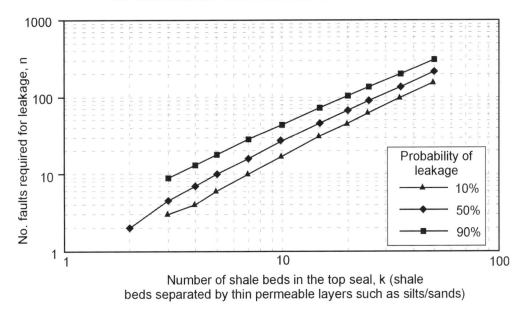

Fig. 7. Probability of top seal leakage. Analytical solution for shale beds of constant thickness, t, in which identical faults of maximum throw, T_{max}, are randomly dispersed.

Discussion

Basin-scale prediction of trap integrity

The power of the techniques outlined above, to predict traps with high leak risk, may be fully realized only if they are applied basin-wide. The simplest way to achieve this is to construct attribute grid maps that highlight the areas of high and low trap integrity. The attributes may be grouped under the headings rheology, fault density and stratigraphy.

Rheology. Mudrock rheology will change over the full extent of a basin, from areas in which rocks are strong and brittle (in uplifted over-consolidated blocks, or locally cemented pockets) to areas in which the rocks are soft and ductile (uncemented, undercompacted, high swelling clay mudrocks). Top-seal integrity relies upon a mudrock possessing ductility, so that it remains sealing after tectonic deformation. Grid maps of rheology may be derived from existing well data (OCR, BRI, friction angle) or, where no well data exist, from seismic interval velocities, to highlight areas in the basin with reduced seal risk.

Stratigraphy. It is extremely valuable to know how the internal stratigraphy of a mudrock formation varies over a basin, because seals in which large amounts of leaky strata, such as coarse silts or sands, exist will be much more prone to leakage in the presence of faults than massive mudrocks. Sub-seismic fault networks will be likely to exploit these leaky strata in a seal via small-scale across-fault juxtapositions. In combination with the rheology information, an attribute map derived from stratigraphic analysis of a regional seal can identify areas in a basin where these juxtaposition leak paths are most likely to exist. Brittle rocks with high net to gross ratios represent high-risk top seals in terms of effective fault- and fracture-induced leak paths.

Fault density. Where high densities of faults exist alongside brittle, relatively high net/gross, top-seal mudrocks, there will be an increased risk that a permeable network of dilatant faults and fractures will exist, leading to loss of hydrocarbons from a trap. Fault trace maps can be converted to fault density maps for prospect screening purposes and if fractures are assumed to cluster closely around larger faults, the fault density map may be interpreted also in terms of the fracture density. Additional qualification of likely leak areas in the regional seal can be achieved if present-day stress orientations are known. If the stress field is known, then specific orientations of faults with low implied normal stress can be highlighted and mapped in a similar way.

Conclusions

Mudrock strength is the most important factor in the determination of leak risk as a result of connected dilatant fractures through a seal. The most practically available variables determining mudrock strength are uplift, compressive- and shear-wave velocities, and surface area. Rocks which have undergone substantial uplift are likely to be overconsolidated and therefore be anomalously strong and prone to the formation of dilatant fractures. A quantitative criterion for mudrock embrittlement, using sonic velocities, has been developed using the brittleness index $BRI = UCS/UCS_{NC}$. For $BRI > 2$ the risk of embrittlement increases with increasing BRI. The surface area of a mudrock correlates with friction angle and therefore its strength. When the derived friction angle of a mudrock is used in combination with sonic data from the same formation, the *in situ* strength may be determined.

In layered clastic top seals, differential leakage is a mechanism that may occur as a result of the different leak characteristics of gas and oil phases in contact with leaky strata within a seal, e.g. fine sandstone. Leakage in such layered top seals will be enhanced by linkage of leaky layers via juxtapositions across sub-seismic faults to form permeable leak pathways. As a rule of thumb, seal leak risk has been calculated as 90% if the number of faults with throws in excess of the mean component mudrock layer thickness in a top seal equals five times the number of mudrock layers comprising the seal.

The techniques briefly outlined here can be enhanced by integration of rheological, structural, stratigraphic and *in situ* stress data. This is most practically achieved by creating basin-wide attribute maps, which can, in combination, separate areas of high seal integrity from areas of high leak risk.

We thank Shell International Exploration and Production for permission to publish work contributing to this paper. We thank D. Olgaard, L. Dell'Angelo and R. Nüesch (ETH Zurich), and M. Fischer (Woodside Energy Ltd) for many stimulating discussions. W. England and S. Hay are thanked for helpful reviews.

References

BEHRMANN, J. H. 1991. Conditions for hydrofracture and the fluid permeability of accretionary wedges. *Earth and Planetary Science Letters*, **107**, 550–558.

DOWNEY, M. W. 1984. Evaluating seals for hydrocarbon accumulations. *Bulletin, American Association of Petroleum Geologists*, **68**, 1752–1763.

DU ROUCHET, J. 1981. Stress fields, a key to oil migration. *Bulletin, American Association of Petroleum Geologists*, **65**, 74–85.

GAARENSTROOM, L., TROMP, R. A. J., DE JONG, M. C. & BRANDENBURG, A. M. 1993. Overpressures in the Central North Sea: implications for trap integrity and drilling safety. *In*: PARKER, J. R. (ed.) *Petroleum Geology of Northwest Europe: Proceedings of the 4th Conference*. Geological Society, London, 1305–1313.

GAUTHIER, B. D. M. & LAKE, S. D. 1993. Probabilistic modelling of faults below the limit of seismic resolution in the Pelican Field, North Sea, offshore United Kingdom. *Bulletin, American Association of Petroleum Geologists*, **77**, 761–777.

GRUNAU, H. R. 1987. A worldwide look at the cap-rock problem. *Journal of Petroleum Geology*, **10**(3), 245–266.

HALL, P. L., MILDNER, D. F. R. & BORST, R. L. 1986. Small-angle scattering studies of the pore spaces of shaly rocks. *Journal of Geophysical Research*, **91**(B2), 2183–2192.

HORSEMAN, S. T., MCCANN, D. M., MCEWEN, T. J. & BRIGHTMAN, M. A. 1986. Determination of the geotechnical properties of mudrocks from geophysical logging of the Harwell boreholes. Report, Fluid Processes Research Group, British Geological Survey, **FLPU 84-14**.

INGRAM, G. M., URAI, J. L. & NAYLOR, M. A. 1997. Sealing processes and top seal assessment. *In*: MOLLER-PEDERSEN, P. & KOESTLER, A. G. (eds) *Hydrocarbon Seals: Importance for Exploration and Production*. Norwegian Petroleum Society (NPF) Special Publication, **7**, 165–175.

MANDL, G. & HARKNESS, R. M. 1987. Hydrocarbon migration by hydraulic fracturing. *In*: JONES, M. E. & PRESTON, R. M. F. (eds) *Deformation of Sediments and Sedimentary Rocks*. Geological Society, London, 39–53.

OLGAARD, D. L., URAI, J. L., DELL'ANGELO, L. N., NUESCH, R. & INGRAM, G. M. 1997. The influence of swelling clays on the deformation of mudrocks. *International Journal of Rock Mechanics and Mining Science*, **34**(3–4).

SCHLOMER, S. & KROOSS, B. M. 1997. Experimental characterisation of the hydrocarbon sealing efficiency of cap rocks. *Marine and Petroleum Geology*, **14**, 565–580.

SCHOWALTER, T. T. 1981. Prediction of cap rock seal capacity. *Bulletin, American Association of Petroleum Geologists*, **65**, 987.

SIBSON, R. H. 1981. Controls on low-stress hydrofracture dilatancy in thrust, wrench and normal fault terrains. *Nature*, **289**, 665–667.

SMITH, R. D. A. 1987. Early diagenetic phosphate cements in a turbidite basin. *In*: MARSHALL, J. M. (ed.) *Diagenesis of Sedimentary Sequences*. Geological Society, London, Special Publications, **36**, 141–156.

STEIGER, R. P. & LUENG, P. K. 1988. Quantitative determination of the mechanical properties of mudrocks. *Society of Petroleum Engineers of AIME*, 18024.

URAI, J. L. 1995. Brittle and ductile deformation of mudrocks. *EOS Transactions, American Geophysical Union*, November 7 1995, F565.

—— & WONG, S. W. 1994. Deformation mechanisms in experimentally deformed shales. *Annales Geophysicae*, **12**(Supplement 1), C98.

——, VAN OORT, E. & VAN DER ZEE, W. 1997. Correlations to predict the mechanical properties of mudrocks from wireline logs and drill cuttings. *Annales Geophysicae*, **15**(Supplement 1), C141.

VAN OORT, E., HALE, A. H., MODY, F. K. & SANJIT, R. 1996. Transport in shales and the design of improved water-based shale drilling fluids. *Society of Petroleum Engineers (SPE) Drilling and Completions, September 1996* (originally presented at 1994 SPE Annual Technical Conference and Exhibition, New Orleans, 25–28 September).

WATTS, N. L. 1987. Theoretical aspects of cap-rock and fault seals for single- and two-phase hydrocarbon columns. *Marine and Petroleum Geology*, **4**, 274–307.

WOOD, D. M. 1990. *Soil Behaviour and Critical State Soil Mechanics*. Cambridge University Press, Cambridge.

Origin of overpressures on the Halten Terrace, offshore mid-Norway: the potential role of mechanical compaction, pressure transfer and stress

T. SKAR[1], R. T. VAN BALEN[1], L. ARNESEN[2] & S. CLOETINGH[1]

[1]*Vrije Universiteit, Faculty of Earth Sciences, De Boelelaan 1085,*
1081 HV Amsterdam, Netherlands
[2]*Norsk Hydro Research Centre a.s., 5020 Bergen, Norway*

Abstract: The Halten Terrace, a structural element on the mid-Norwegian margin, is characterized by distinct spatial variations in subsurface fluid pressure. Jurassic formations on the western part of the Halten Terrace are highly overpressured (>30 MPa), whereas the same formations have fluid pressures close to hydrostatic in the central and eastern parts of the terrace. The Cretaceous formations are moderately overpressured in the whole area. The porosities in the overpressured middle Jurassic formations and lower Cretaceous shales are only slightly influenced by the present-day overpressures, suggesting that most of the overpressure development was recent. Overpressure generation as a result of mechanical compaction, induced by the weight of the Plio-Pleistocene sediments, is not sufficient to explain the overpressure magnitudes in the Jurassic and Cretaceous formations. Results from forward modelling of mechanical compaction suggest that the deep Rås Basin adjacent to the Halten Terrace has been highly overpressured since Cretaceous time. Pressure transfer from this basin to the western part of the Halten Terrace is therefore a possible explanation for the observed overpressures. Migration of the overpressured fluids may have occurred through the Klakk Fault Complex separating the basin from the terrace, during an event related to a basinal stress reorientation in early Pliocene time.

In the petroleum industry, knowledge of the distribution of overpressures in sedimentary basins is critical in exploration, drilling and production operations (Fertl *et al.* 1994). However, despite the need to understand the controls on subsurface fluid pressures, there has been much disagreement on the origin of overpressures. According to Osborne & Swarbrick (1997), overpressure can be produced by (1) an increase in compressive stress (compaction disequilibrium and tectonic stress), (2) changes in the volume of the pore fluid or rock matrix (e.g. hydrocarbon generation, aquathermal expansion, clay dehydration) and (3) fluid movement and buoyancy. The process of compaction disequilibrium (mechanical compaction) has received by far the most attention and has been applied successfully to explain overpressures in young sediments in rapidly subsiding basins such as the Gulf Coast basins (e.g. Dickinson 1953). The compaction disequilibrium model is based on the assumption that compaction is governed by vertical effective stress, which is defined as the difference between the overburden and the fluid pressure. If fluids are trapped in the compacting sediments during burial the pore fluids will support part of the overburden, thus creating fluid overpressures. According to the compaction disequilibrium model, overpressured formations have higher porosities than their normally pressured equivalents. In contrast, overpressures caused by volume changes, fluid movements and buoyancy can decrease the effective stress without affecting the porosity, a situation often referred to as sediment unloading (e.g. Ward 1995). In this case, overpressured formations can have the same porosity as formations with hydrostatic fluid pressure at a given depth. Thus, examination of the porosity distribution in formations with different fluid pressures can contribute to an increased understanding of the cause of overpressuring.

Lateral fluid flow within sands in a layered sand-shale sequence could be an important component of the compaction process (e.g. Magara 1976; Bredehoeft *et al.* 1988). Thus, fluid movement as a result of compaction can increase fluid pressures in the up-dip end of laterally extensive permeable layers (Mann & Mackenzie 1990). In addition, abnormal fluid pressures may also be due to pressure transfer along fault zones (e.g. Grauls & Baleix 1994).

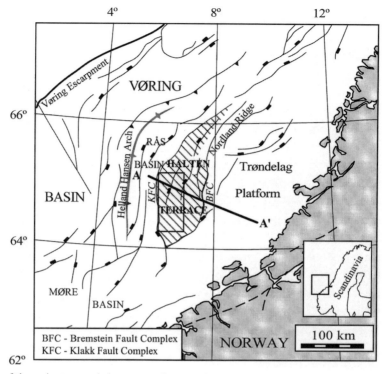

Fig. 1. Map of the main structural elements on the central part of the mid-Norwegian margin. The boxed area marks the location of the study area, which is shown in Fig. 3. The bold line indicates the position of the cross-section in Fig. 2.

The occurrence of fluid movements may have direct implications for prediction of anomalous pressures, secondary migration of hydrocarbons and the efficiency of caprocks. The potential role of mechanical compaction and lateral pressure transfer on overpressure development will be the main issues addressed in this paper.

The study area is located on the Halten Terrace, offshore mid-Norway (Fig. 1), where fluid pressures in the Jurassic and Cretaceous formations are characterized by distinct spatial variations. The Halten Terrace is part of a large mid-Norwegian extensional sedimentary basin which has originated from multiple extensional faulting, particularly during rifting episodes in Mesozoic and Cenozoic time. The last rifting episode that affected this broad area preceded and terminated with the opening of the northern part of the North Atlantic Ocean during earliest Eocene time (Skogseid et al. 1992). From late Eocene time large compressional dome and arch structures developed along the central and western part of the margin, and were reactivated during late Miocene–Pliocene time with accelerated subsidence in basinal areas and uplift of the Norwegian mainland (e.g. Riis 1996).

The Halten Terrace, located in the east central area of the present margin, became a separate structural element mainly during the extensive late Jurassic–early Cretaceous rifting, and was only slightly tectonically affected by the last rifting episode, which terminated with continental break-up. Major faulting separated the terrace from the Trøndelag Platform to the east by the Bremstein Fault Complex and from the deep Rås Basin in the west by the Klakk Fault Complex (Figs 1 and 2). Tilted fault blocks were formed during an early stage of the formation of the Halten Terrace (Blystad et al. 1995; Koch & Heum 1995), with major implications for more recent hydrodynamic systems.

We investigate the overpressure build-up history at the western part of the Halten Terrace, using well logs and forward modelling of gravitational compaction. In the first part of the paper we review the overpressure data and pressure systems, and discuss the results of an analysis of fluid pressure–porosity relationships within the middle Jurassic and lower Cretaceous formations. Next, a model for the overpressure evolution is presented. The model is based on sudden pressure build-up by lateral fluid transfer

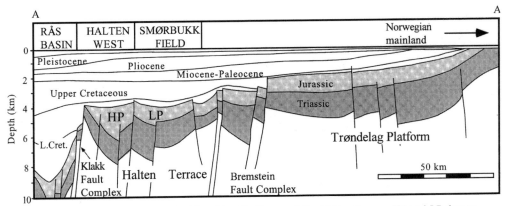

Fig. 2. Simplified cross-section across the Trøndelag Platform and the Halten Terrace. HP and LP denote formations with high fluid overpressures and low fluid overpressures, respectively. (Note the structural relief between the Halten Terrace and the Rås Basin.)

from the nearby deep Rås Basin through the Klakk Fault Complex, which separates the basin from the terrace. In this model the dynamic stress-dependent hydraulic characteristics of the Klakk Fault Complex play a crucial role. Thereafter, a probable timing of the fluid pressure transfer event is proposed. Finally, the potential of other overpressure mechanisms is discussed.

Fluid pressure distribution on the Halten Terrace

Fluid pressure data, pressure systems and hydrocarbon occurrences

Fluid pressure data have been compiled from tests (Repeat Formation Tester; RFT) performed in Jurassic reservoirs, mainly the Garn and Ile Formations, and in the Cretaceous Lysing and Lange sandstones from 23 wells on the Halten Terrace (Fig. 3). The majority of the pressure data are gathered from the middle Jurassic sandstones, as these reservoirs are the main targets for hydrocarbon exploration in the study area. The pressure data are plotted in Fig. 4 and show that the fluid pressures are characterized by significant variations. On the basis of the regional distribution of fluid pressure in the Jurassic reservoirs, three main pressure compartments can be distinguished (Fig. 3). The wells drilled on the western part of the terrace (also referred to as Halten West) have encountered reservoirs with fluid overpressures in the range 20–35 MPa (high-pressure regime). Low fluid overpressures (<7.5 MPa) and hydrostatic conditions prevail in the reservoirs on the central and eastern parts of the Halten Terrace and on the Trøndelag Platform (referred to as the low-pressure regime).

A transitional area with reservoir overpressure 7.5–20 MPa is encountered on the southern part of the Halten Terrace and includes the oil-producing Njord Field (moderate-pressure

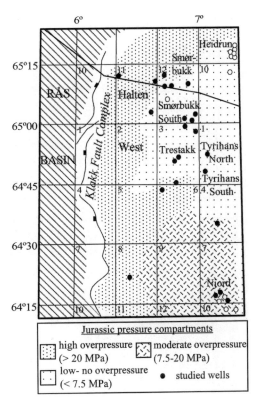

Fig. 3. Location map of the Jurassic pressure compartments and studied wells on the Halten Terrace. The map is modified from Hermanrud et al. (1998). The bold line indicates the location of parts of the cross-section shown in Fig. 2.

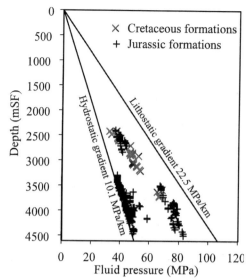

Fig. 4. Compilation of pressure data (from Repeat Formation Tester) for Jurassic and Cretaceous sandstones in the study area. The location of the wells from which pressure data for the Jurassic formations are included is shown in Fig. 3. mSF, meters below sea floor.

diagenetic effects as the cause of the pressure difference cannot be ruled out. In contrast to the fluid pressure patterns in the Jurassic reservoirs, the fluid pressures in the Cretaceous formations show, in general, a uniform distribution with increasing overpressure with depth (Fig. 4).

Figure 5 shows the relationship between well locations, formation burial depths and fluid pressures for the upper and middle Jurassic sequence in the vicinity of the Smørbukk area for three wells that are less than 6 km apart. A fluid pressure difference of 30 MPa is observed between the water-bearing reservoirs in well 6506/11-1 on Halten West and the hydrocarbon-bearing reservoirs in wells 6506/12-1 and 6506/12-6 in the Smørbukk area. The major difference in fluid pressures provides evidence for the existence of two separate present-day hydrodynamic systems in this area.

The main hydrocarbon discoveries are encountered in the Jurassic reservoirs within the low-pressure regime and in the Njord Field (overpressure c. 10 MPa). Reservoirs in the high-pressure regime are mainly devoid of hydrocarbons, although the Jurassic sequence is buried to similar depths and contains the same stratigraphic units as in the adjacent areas where reservoirs are filled with hydrocarbons (Fig. 5). The relationship between high overpressure and water-filled reservoirs has led several researchers to propose a late phase of overpressure

regime). Fluid pressure compartments bounded by faults can best explain the overpressure pattern in the Jurassic formations (e.g. Koch & Heum 1995), although lithological and/or

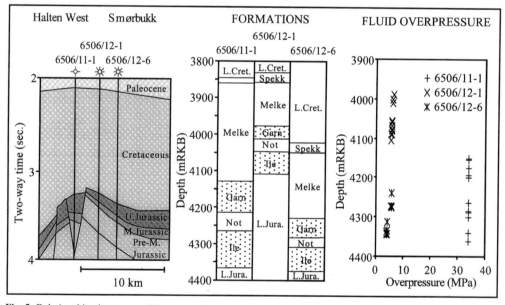

Fig. 5. Relationships between well locations, fluid overpressure and burial depths of middle Jurassic formations. The fluid pressure data are from the Garn and Ile Formations. The cross-section is modified from Ehrenberg (1992) and runs parallel to the cross-section shown in Fig. 3.

development, which caused hydraulic fracturing with subsequent leakage of hydrocarbons (e.g. Ungerer et al. 1990). However, as noted by Hermanrud et al. (1998), hydrocarbons have been encountered in North Sea reservoirs with even higher overpressures than in the reservoirs on Halten West.

Fluid pressures in fine-grained lithologies

The dominant lithology in the post-Jurassic succession on the Halten Terrace is shale. The fluid pressure distribution in these fine-grained sediments is to some extent uncertain, as the fluid pressures are evaluated using drilling data and wireline logs. Estimation of fluid pressure from drilling data (i.e. drilling exponent) and well logs relies on establishing a normal trend line. Deviations from this trend are used for calculation of overpressure. However, deviations from the inferred normal trend line can be due to factors such as lithology changes and mudstone mineralogy, which can be wrongly interpreted as fluid pressure variations (Aplin et al. 1995).

Despite the uncertainties related to determination of fluid pressures in the fine-grained lithologies, most fluid pressure profiles presented in the final well reports show a consistent fluid pressure pattern with depth. The relationship between log responses and a predicted fluid pressure profile on the Halten Terrace is shown in Fig. 6. As seen, the fluid pressure predictions (in equivalent mud weight) are strongly influenced by the log trends, particularly down to 3 km. Fluid pressures within most of the Tertiary succession are predicted to be close to hydrostatic, and they increase rapidly to a mud weight equivalent of 1.5–1.7 g cm^{-3} in the lowermost Tertiary–uppermost Cretaceous deposits corresponding to the anomalous log responses at these depths. The fluid pressure is predicted to remain rather constant throughout the upper Cretaceous shales and to increase gradually in the lower Cretaceous shaly formations. A new increase in fluid pressure is usually predicted within the lowermost Cretaceous and upper Jurassic shales. The fluid pressure in the Jurassic reservoirs increases further to a mud equivalent of 1.7–1.85 g cm^{-3} on the westernmost part of the Halten Terrace, whereas it is close to hydrostatic in the Smørbukk Field (see Fig. 5).

Fluid pressure–porosity relationships

Information from the density and sonic logs, measured sandstone porosities and fluid pressure data were compiled to investigate the influence of fluid pressure on the porosity distribution in the middle Jurassic and lower Cretaceous formations. Porosities for the fine-grained intervals have been derived from the sonic and density logs. Porosity was calculated from bulk density data by applying the relationship

$$\phi = (\rho_{\text{matrix}} - \rho_{\text{bulk}})/(\rho_{\text{matrix}} - \rho_{\text{fluid}})$$

where a constant shale grain density of 2.72 and a fluid density of 1.05 is assumed. Sonic porosities for the Cretaceous shales were derived using the relationship between sonic transit times and porosity for Cretaceous and Tertiary shales on the Norwegian shelf as proposed by Hansen (1996):

$$\phi = (1/1.57)[(\Delta t - 59)/(189 - 59)].$$

The porosity distribution in the middle Jurassic formations is expected to show lateral variations as a result of the different pressure regimes. We have analysed the porosity distribution for three formations located within the high-pressure and low-pressure regimes: the Garn Formation (sandstone), Ile Formation (sandstone) and Not Formation (fine-grained lithology). Porosity values for the sandstones are from core measurements, whereas porosities for the Not Formation are derived from the logs. Figure 7a shows sandstone porosities in the Garn Formation from 13 wells, of which four wells are located in the high-pressure regime. The highly overpressured sandstones have porosities that are on average 3–5% higher than the porosities from the slightly overpressured sandstones. A similar porosity difference is encountered for the Ile Formation (Fig. 7b). These observations will be discussed in more detail below.

Bulk density and interval transit times for the fine-grained Not Formation are shown in Fig. 8. The bulk density data show almost no difference between the two pressure regimes (it is assumed that the fluid pressure in the Not Formation is the same as in the overlying Garn Formation), whereas the interval transit times are significantly higher in the high fluid pressure regime compared with the low-pressure regime. To better illustrate the differences between the sonic and bulk density responses, the log data have been converted to porosity units. The density-porosities are plotted in Fig. 9a. The data show that the porosities on Halten West are on average 1–3% higher than in the low-pressure regime. In contrast, sonic-porosities on Halten West are significantly higher than their equivalents in the low-pressure regime (Fig. 9b). The sonic-porosities are 5.5–9% higher on Halten West depending on which relationship is used for

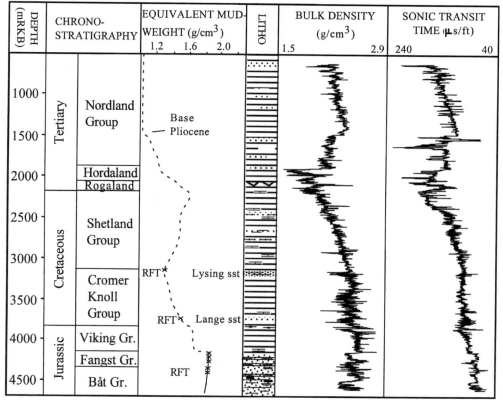

Fig. 6. Typical bulk density and sonic transit time responses in wells on Halten West. The fluid pressures (in equivalent mud weight) is based on evaluation of drilling data and wireline logs. RFT denotes where pressure measurements in the Jurassic and Cretaceous sandstones are usually available.

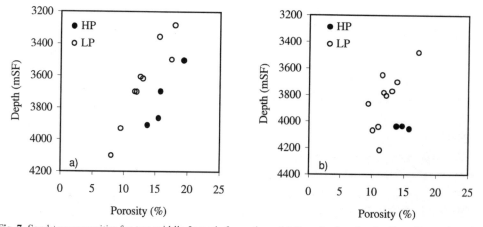

Fig. 7. Sandstone porosities for two middle Jurassic formations. (**a**) Porosity data for the Garn Formation (from Ehrenberg, 1990); (**b**) Routine core He-porosities for the Ile Formation. HP and LP denote formations with high fluid overpressure (>25 MPa) and low fluid overpressure (<7.5 MPa), respectively. Depth are in metres below sea floor.

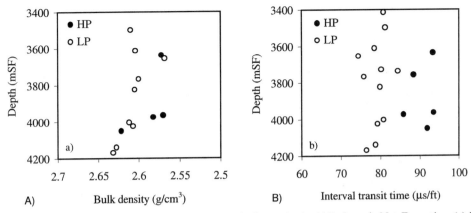

Fig. 8. Bulk density and interval transit time data for the fine-grained middle Jurassic Not Formation. (**a**) Bulk densities; (**b**) interval transit times. HP and LP denote formations with high fluid overpressure (>25 MPa) and low fluid overpressure (<7.5 MPa), respectively. mSF, metres below sea floor.

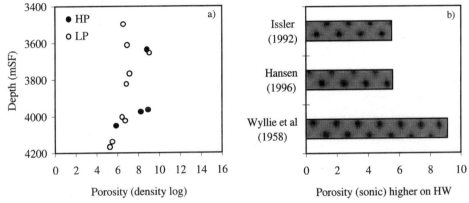

Fig. 9. (**a**) Porosities derived from the bulk density data in Fig. 8a. (**b**) The average difference in sonic-porosities for the highly overpressured formations and normally pressured formations using three different interval transit time–porosity equations for calculations of sonic-porosity. The calculations are based on the interval transit times in Fig. 8b. mSF, metres below sea floor.

determination of porosity from the sonic log. Thus, the log data illustrate that the sonic-porosities on Halten West are much higher than in the Smørbukk area, whereas the bulk density data indicate that there are only minor differences in porosities within the two pressure regimes. Similar observations were reported by Hermanrud et al. (1998), who interpreted the sonic log to be influenced by fluid overpressuring (as a result of reduced effective stress), which caused increased travel times. We agree with the interpretation of Hermanrud et al. (1998) in that the sonic log is unreliable as a porosity indicator in highly overpressured formations and, accordingly, suggest that the density log is more reliable for porosity predictions. For a more detailed discussion related to factors that may have influenced the sonic and density log readings in the Not Formation, the reader is referred to Hermanrud et al. (1998).

The fluid pressure distribution in the Cretaceous formations is partly constrained by RFT measurements at depths around 2.9–3.1 km (Lysing Formation) and 3.5–3.7 km (Lange Formation). The overpressure data plotted in Fig. 4 show a gradual increase in overpressure with depth from c. 14 MPa at c. 3 km to 26 MPa at around 3.7 km. We have analysed the log-porosity trends in the lower Cretaceous shales for seven wells in the Smørbukk area to examine whether this overpressure increase is detected in the data sets. The shales used in the analysis have been identified from the gamma log, separation of the neutron and density logs, and from lithological descriptions in the completion reports. They are averaged over 10 m intervals.

The trends of the density-porosities and sonic-porosities are shown in Fig. 10. The density-porosities have rather low values (<15%) below c. 3 km depth and are reduced at a rate of c. 10% km^{-1}. The reduction in sonic-porosities over the same depth interval is much less (c. 5% km^{-1}). The different depth trends for the density-derived and sonic-derived porosities can be visualized by comparing the obtained trends with the exponential normal compaction curve proposed by Hansen (1996) for Tertiary and Cretaceous shales on the Norwegian shelf. (It should be noted that the curve is based on data points down to c. 2.6 km. This curve is extrapolated to greater depths as no other normal compaction curves exist for the Cretaceous shales on the mid-Norwegian margin.) On the basis of the assumptions for calculations of porosity, there is a pronounced difference in the porosity–fluid pressure relationships. The density-porosities decrease gradually with depth and follow a trend that is almost parallel to the 'normal compaction curve'. The sonic-porosities, however, deviate from the normal trend with depth. Assuming that the density log is the most reliable porosity indicator, the behaviour of the sonic-porosities with depth may reflect the fact that the sonic log readings are influenced by an increase in overpressure (as in the case with the Not Formation). Alternatively, the trend of the sonic-porosities can be attributed to variations in the sediment composition (mineralogy, texture), which may influence the sonic log readings (Vernik & Nur 1992; Luo & Vasseur 1993; Vernik 1997).

In summary, formation porosities in the Not Formation on Halten West are apparently only slightly influenced by the present-day high overpressure. Determination of porosity from well logs is, however, associated with uncertainties. On the basis of the bulk density data, shale porosities are on average 1–3% higher in the high-pressure regime than in the low-pressure regime. Errors in the data set can probably be a few porosity per cent, which suggests that the difference in porosity that may be due to overpressure is in the range 0–5%. From this inference, three different end-member interpretations can be made: (1) porosities in the Not Formation are not influenced by the present-day overpressures, which suggests that shale compaction may proceed without hindrance by overpressure (Hermanrud et al. 1998); (2) overpressuring is recent and post-dates compaction; (3) the Not Formation is undercompacted, which reflects hampering of the compaction process as a result of overpressure.

The sandstone porosities on Halten West are higher than in the low-pressure regime (3–5%; see Fig. 7). Although sandstone porosity at greater depths on the Halten Terrace is strongly controlled by cementation processes (e.g. Ehrenberg 1990; Bjørlykke & Egeberg 1993), there are strong indications that the porosity difference is due to fluid overpressure. Ehrenberg (1990) examined factors controlling the porosity distribution in the Garn Formation and found that there is no correlation between high porosity and

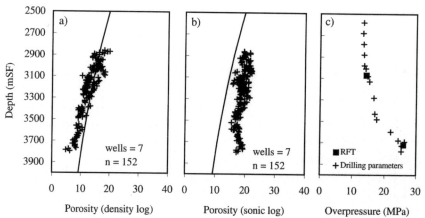

Fig. 10. Relationship between log-derived porosities and fluid overpressures in the lower Cretaceous shaly formations in the Smørbukk area. (a) Shale porosities derived from the bulk density data; (b) shale porosities derived from the sonic data; (c) RFT measurements and fluid pressure estimates based on drilling data from well 6506/11-2. Each porosity value is an average over a 10 m interval. The continuous curve in the porosity–depth plots is the 'normal compaction curve' proposed by Hansen (1996) for Tertiary and Cretaceous shales on the Norwegian margin. (See text for discussion.)

grain size, sorting and primary mineralogical composition (feldspar and mica). Ehrenberg (1990) concluded that overpressuring appears to be favourable for porosity preservation in the Garn Formation below 3 km burial depth, resulting in about 3% higher porosity than for normally pressured reservoirs. Moreover, Ramm & Bjørlykke (1994) examined porosity–depth trends for 260 cored sandstone units from 110 wells from the North Sea and Haltenbanken area and assessed the quantitative effects of varying fluid pressure, temperature history and mineralogy on the porosity distribution. They showed that sandstone porosities at a given burial depth are higher within highly overpressured sandstones (pressure gradients $>1.6–1.7$ g cm^{-3}) relative to normally pressured sandstones and concluded that the amount of chemical compaction is reduced because of the high overpressure.

The lower Cretaceous shales have formation porosities that are generally low ($<15\%$ below 3 km depth). As the Cretaceous formations are overpressured throughout the study area, there are no accurate curves that describe the compactional behaviour of normally compacted shales. However, by comparing the porosities derived from the bulk density data with the normal compaction curve of Hansen (1996), it appears that the shales are close to normally compacted rather than severely undercompacted.

In this paper, we examine the scenario that the middle Jurassic formations are slightly undercompacted. The other alternatives will be briefly discussed in the 'Discussion and conclusions' section. In the following section, we investigate the role of compaction disequilibrium in the evolution of overpressure on the Halten Terrace. First, the role of mechanical loading caused by the weight of the Plio-Pleistocene sediments is evaluated using simple theoretical considerations. We then apply a 1D numerical model to predict the differences in porosities that will occur if overpressures are caused by sediment loading. Thereafter, we predict (using the 1D model) overpressure and porosity evolution in the nearby deep Rås Basin and propose a model that includes the transfer of overpressure from this basin and into the Halten Terrace area.

Compaction disequilibrium

The temporal evolution of fluid pressure as a result of mechanical compaction is controlled by the relationship between sediment permeability and sedimentation rates (e.g. Audet & McConnell 1992). Sedimentation rates are constrained from well data and provide a first-order indication of when the most pronounced periods of overpressure build-up because of sediment loading may have occurred. Sedimentation rates (not corrected for compaction) on the Halten Terrace are depicted in Fig. 11 and it can be seen that there are two periods with high sedimentation rates; from 92 to 84 Ma and during Plio-Pleistocene times. A starting point is therefore to evaluate the effect of Plio-Pleistocene loading on overpressure development in the Cretaceous and Jurassic formations. These predictions can then be related to the observed porosity trends (e.g. in the Not Formation) where the amount of undercompaction as a result of sediment loading can be predicted. For the porosity prediction, we have used a forward model that simulates the dynamic porosity and fluid pressure evolution as a result of mechanical compaction.

The role of Plio-Pleistocene loading on overpressure development

The amount of overpressure generated by the Plio-Pleistocene sediment loading is related to the thickness of the Plio-Pleistocene deposits, which is about 1.2 km on the western Halten Terrace. Theoretically, the maximum amount of fluid pressure increase is limited by the increase in lithostatic pressure, imposed by the weight of the Plio-Pleistocene sediments. For the deposition of a 1.2 km thick sequence of sediments, the lithostatic pressure increase in the underlying sediments is given by

$$\Delta P_{\text{lith}} = g \times \Delta z \times [\phi \rho_{\text{fl}} + (1-\phi)\rho_{\text{grain}}]$$
$$9.8 \times 1200 \times (0.3 \times 1000 + 0.7 \times 2700)$$
$$= 26.3 \text{ MPa}$$

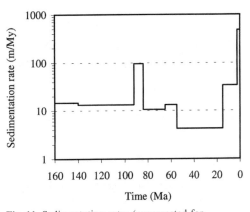

Fig. 11. Sedimentation rates (uncorrected for compaction) based on chronostratigraphic information from well 6506/11-1 located on the western part of the Halten Terrace.

where g is the gravitational acceleration, Δz is the thickness of Plio-Pleistocene deposits, ϕ is the porosity (here taken as 30%), ρ_{fl} is the density of water and ρ_{grain} is the grain density.

In undrained conditions (no inflow or outflow) the total fluid pressure increase below the sediment load as a result of the lithostatic pressure increase is given by

$$\Delta P_{fl} = B \times \Delta P_{lith}$$

where B is the Bishop-Skempton–Biot pore pressure build-up coefficient, a porosity-dependent material property (Palciauskas & Domenico 1982; Green & Wang 1986). Measured B values for sediments range from 0.99 to about 0.6. In general, weak, uncompacted sediments have a B value close to unity. With continuing compaction sediments become mechanically stronger and their B value decreases correspondingly.

The increase in overpressure equals the increase in fluid pressure minus the hydrostatic pressure increase:

$$\Delta P_{ov} = \Delta P_{fl} - \rho_{fl} g \Delta Z$$

where it is assumed that the amount of subsidence of the compacting sediment equals the thickness of the Plio-Pleistocene sediment load. Using an 'average' B equal to 0.85, the overpressure increases as a result of loading in the sediments underlying the Plio-Pleistocene wedge is:

$$\Delta P_{ov} = (0.85 \times 26.3) - (9.8 \times 1 \times 1.2)$$
$$= 10.6 \text{ MPa}.$$

Sediments with $B \approx 1$ would experience an overpressure increase of about 14.5 MPa, which is the maximum obtainable overpressure in the area of interest. These inferences are based on undrained conditions. For drained circumstances the increase in overpressure would be less because of fluid loss. The expected maximum increase in overpressure in the Jurassic and Cretaceous sequences as a result of the Plio-Pleistocene sedimentation is, therefore, less than 11 MPa. As a result, compaction disequilibrium because of the Plio-Pleistocene loading fails to explain the overpressures measured in the upper Cretaceous Lysing Formation (>14 MPa), lower Cretaceous Lange Formation (>20 MPa) and in the Jurassic formations (>30 MPa).

This inference can be further tested by comparison of observed and modelled porosity. If overpressures were generated as a result of sediment loading, the expected porosity differences between normally compacted and overpressured formations (in this case, the middle Jurassic Not Formation) can be predicted using a forward model of mechanical compaction.

The numerical model and input data

A 1D, single-phase fluid model is used to investigate the influence of fluid pressures on porosity in a compacting sedimentary sequence. The model is based on three assumptions: (1) Darcian fluid flow in the porous sediment; (2) Terzaghi's principle of effective stress; (3) a constitutive law for deformation of the rock frame expressed as a porosity–effective stress relationship. These three assumptions are combined with mass conservation of the solid and fluid phases. The differential equations for excess pressure generation are solved using a finite element scheme. For a full description of the forward model, the reader is referred to Kooi (1997).

The geological data needed for modelling of the dynamic fluid pressure and porosity evolution are a lithological column, chronostratigraphic information and equations describing porosity and permeability reduction with effective stress for each lithology present in the column. Each lithology is characterized by: (1) ρ_{gr}; (2) $\phi = \phi_0 \exp(-\beta \sigma_{eff})$; (3) $k = 10^{-a+\phi b}$, where ρ_{gr} is grain density, ϕ is porosity, ϕ_0 is surface porosity, β is bulk compressibility, σ_{eff} is effective stress, k is permeability, and a and b are constants.

Predictions of relative porosity differences as a result of fluid overpressure

Predictions of the amount of undercompaction (expressed as the porosity difference between normally compacted and overpressured sediments) rely on information about the porosity–depth trends for normally compacted sediments. Hermanrud et al. (1998) proposed a normal compaction curve for the Not Formation (on the basis of porosities derived from the density logs) expressed as $\phi = 58.93 \, e^{-0.00051z}$. For the Cretaceous formations, the compaction curve of Hansen (1996), described as $\phi = 71 \, e^{-0.00051z}$, is used as a 'reference curve'. In the modelling, it is assumed that porosity is an exponential function of effective stress, where the normal trend line is determined by the two input variables ϕ_0 and β. The curve for the Not Formation can be approximated by $\phi_0 = 0.5$ and $\beta = 0.039$, whereas the curve for the Cretaceous shales can be approximated by $\phi_0 = 0.55$ and $\beta = 0.038$. The compaction curves and the normal compaction curves predicted by the model deviate above c. 1.5 km (Fig. 12a), but this does not influence our predictions as we intend only to predict the porosity differences at depths below c. 3 km, from where the curves almost coincide. To examine how the porosities are influenced by an increase in fluid overpressure, we have used several assumptions for shale permeability, as this variable is the main control on overpressure development. The input data and additional values of constants for the modelling are shown in Tables 1 and 2.

Figure 12b shows the predicted undercompaction. The Not Formation is buried to depths between 3400 and 4200 m. If the maximum difference in porosity between the overpressured and normally pressured shales of the Not Formation is five porosity units, then the amount of undercompaction as a result of sediment loading corresponds to overpressures of c. 14 MPa at 4200 m depth and 12 MPa at 3500 m depth (Fig. 12b). The difference in overpressure magnitudes required to produce the same amount of undercompaction is because the rate of porosity reduction for a normally pressured sediment at shallow depth is higher than during deeper burial, as the porosity–depth curve is assumed to be exponential. Thus, the predicted amount of undercompaction is higher at shallow depths than at larger depths for a given magnitude of overpressure. These porosity–fluid overpressure predictions indicate

Table 1. *Values of constants*

μ	5×10^{-4} Pa s
β_{fl}	4.3×10^{-10} Pa^{-1}
ρ_{fl}	1030 kg/m^{-3}
g	9.8 m s^{-2}
ρ_{gr} (shale)	2720 m s^{-3}
ρ_{gr} (sand)	2650 m s^{-3}

that the present-day observed overpressures have developed late, as overpressures inherited from pre-Pliocene times would have resulted in an even higher amount of undercompaction. Thus, on the basis of modelling predictions, compaction disequilibrium is insufficient to explain the present-day high overpressures, and an additional source(s) as the cause of these excess pressures must be considered.

Lateral pressure transfer

Lateral pressure transfer may have played a key role in the evolution of overpressures on the Halten Terrace, as the highest observed overpressures on the terrace are located adjacent to the deep Rås Basin (Fig. 3). The Rås Basin is located between 64°N and 66°40′E, and 4°30′N and 7° 20′E, and has an areal extent that is several times larger than the western part of the Halten Terrace. The Rås Basin is separated from the Halten Terrace by the Klakk Fault Complex,

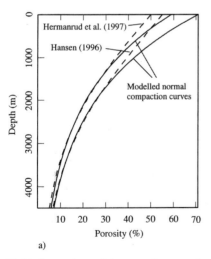
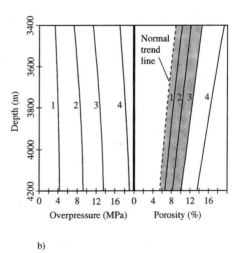

Fig. 12. (a) Comparison of the normal compaction curves proposed for the Not Formation (Hermanrud *et al.* 1998) and for the Cretaceous shales (Hansen 1996) with the normal compaction curves predicted by the model. (b) Modelling predictions of porosity–fluid overpressure relationships for the Not Formation. The grey-shaded area indicates the possible amount of undercompaction when the porosity difference between normally and overpressured shales is in the range 0–5%. The numbers refer to the curves predicted by the model for different assumptions of shale permeability. (See text for discussion.)

Table 2. *Geological information used as input for modelling of the dynamic fluid pressure and porosity evolution shown in Figs 12, 14 and 15*

Thickness (m)		Lithology	Lithology code		Duration on (Ma)
HW	Rås		Fig. 12	Figs 14 & 15	
1200	1200	silt	1	1	2.5
400	400	silt	1	1	12.5
250	250	shale	5	2	40
150	150	shale	5	2	10
200	300	shale	5	2	19
250	900	shale	5	2	3
350	1000	shale	5	2	5
700	4000	shale	5	2	48
300	300	shale	5	3	20
100	100	Garn Fm (sst)	5	4	5
50	50	shale	5	3	5
100	100	Ile Fm (sst)	5	4	5
50	50	shale	5	3	5

Lithology code	ϕ_0	β (MPa^{-1})	a	b
1	0.4	4.0×10^{-2}	15.8	8
2	0.55	3.8×10^{-2}	21.4	8
3	0.5	3.9×10^{-2}	21.4	8
4	0.45	2.3×10^{-2}	15.8	8
5 (shale)	0.50	3.9×10^{-2}	15.8–21.8	8

Each lithology is assigned values for surface porosity (ϕ_0), bulk compressibility (β) and the constants a and b, which determine permeability reduction with effective stress. The two columns for thickness represent the input from Halten West (HW) and the Rås Basin. The two lithology code columns represent the input used to construct Figs 12, 14 and 15.

and subsided rapidly during the late Jurassic–early Cretaceous rifting. The high rate of subsidence created accommodation space for deposition of thick layers of Cretaceous shales. Burial diagrams showing the different subsidence histories for the two areas are shown in Fig. 13. The middle Jurassic sequence is buried to more than 9 km in the Rås Basin, whereas the same stratigraphic interval on Halten West is buried to depths of about 4–6 km. The deeper part of the Rås Basin is therefore considered to be a potential source of high overpressure. One might, therefore, speculate that migration of overpressured fluids from the deep basin into the Halten Terrace area may have occurred through weak spots in the Klakk Fault Complex.

Overpressure development in the Rås Basin

The overpressure evolution in the Rås Basin is modelled using the 1D numerical model described in the previous section. No wells have been drilled in this basin and, therefore, the modelled column represents a synthetic well from the axial part of the basin. The lithostratigraphic and chronostratigraphic input data are constrained from seismic profiles, information on depositional ages (Hagevang and Tørudbakken 1996) and lithology (Dalland et al. 1988). As we intend to illustrate the relative difference in overpressure build-up that may have occurred in the Rås Basin and on the western part of the Halten Terrace, the overpressure evolutions for both areas are predicted using the same assumptions for porosity and permeability reductions. Olstad et al. (1997) presented shale permeability data for the upper Jurassic Melke Formation ranging from 10^{-20} to 10^{-21} m^2. In the modelling, we have therefore adopted 10^{-21} m^2 as a lower limit for shale permeability, where it is assumed that the permeability is reduced four order of magnitude from surface conditions. The input data for the modelling are summarized in Table 2.

The predicted fluid overpressure development through time for the mid-Jurassic interval in the Rås Basin and on Halten West for similar permeability assumptions is illustrated in Fig. 14. The model predicts a pronounced pressure build-up in the Rås Basin throughout early and early late Cretaceous time as a result of the high sedimentation rates that prevailed

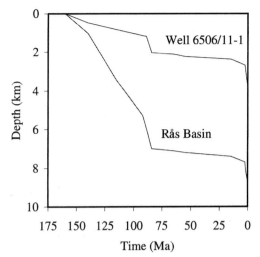

Fig. 13. Burial diagrams (corrected for compaction) showing the subsidence history for the axial parts of the Rås Basin and on Halten West (well 6506/11-1).

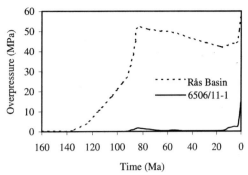

Fig. 14. Modelled 1D overpressure evolution through time for the middle Jurassic interval in the Rås Basin and on the western part of the Halten Terrace. (See text for discussion.)

during this period. These overpressures decline slightly during the end of the Cretaceous period and throughout most of Tertiary time. The predicted overpressure magnitude in the deepest part of the basin throughout Tertiary time is in excess of 42 MPa, which is higher than the present-day overpressures on the Halten Terrace. In contrast, the model predicts that the fluid pressure on the Halten Terrace is close to hydrostatic before Plio-Pleistocene time. A final period of pressure build-up in both regions is associated with the progradation and deposition of a thick pile of clastic sediments during Plio-Pleistocene time. The modelled fluid pressure and porosity profiles from the Rås Basin and Halten Terrace at 2.5 Ma (onset of Plio-Pleistocene subsidence) are shown in Fig. 15. Thus, the modelling results indicate that fluid pressures are close to hydrostatic on the Halten Terrace before Plio-Pleistocene time, whereas the deep Rås Basin is highly overpressured for the given assumptions of shale permeability.

The magnitude of excess fluid pressures that can have developed in the Rås Basin is limited by the formation strength of the seal. During Tertiary time, the Jurassic formations were buried to around 7–8 km and the fluid pressure–lithostatic pressure ratio in the caprock was modelled to be <0.7, which is below the fracture pressure suggested by Gaarenstroom et al. (1993) for deeper parts of the Central North Sea Graben. These modelling predictions suggest that expulsion of overpressured fluids as a result of hydraulic fracturing is not likely to have occurred. Thus, on the basis of our assumptions of shale permeabilities it can be concluded that formations in the deep part of the Rås Basin are likely to have been highly overpressured during Tertiary times. Migration of these overpressured fluids into the Halten Terrace area in sub-Recent times may therefore be sufficient to explain that part of the high overpressures in the Jurassic formations that is not due to the loading of the Plio-Pleistocene sediments. However, as this modelling is one dimensional some complementary calculations of inflow and outflow of fluid volumes and the influence on pressure increase (on Halten West) and dissipation (in Rås Basin) are required.

Fluid volume calculations

Migration of fluid volumes because of differences in overpressure can be treated analytically using the principle of mass conservation. The equation of state for the fluid gives the density change as a result of a change in pore fluid pressure:

$$\partial \rho_{fl} = \rho_{fl} \beta \partial P_{fl}. \qquad (1)$$

The present-day density of the pore fluids at the western part of the Halten Terrace can be expressed as the sum of the fluid density before the overpressure change and the fluid density increase as a result of the increase in overpressure:

$$\rho_{fl(today)} = \rho_{fl(before)} + \partial \rho_{fl}.$$

Using equation (1) gives

$$\rho_{fl(today)} = \rho_{fl(before)} + \rho_{fl(before)} \beta \partial P_{fl}. \qquad (2)$$

Mass balancing requires that the present-day fluid mass at Halten West equals the sum of the

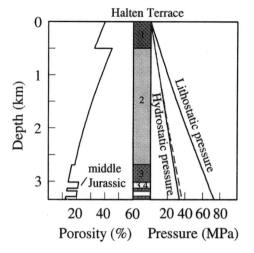

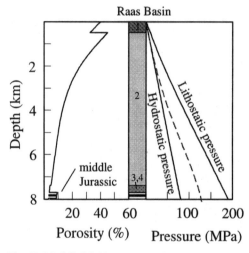

Fig. 15. Modelled fluid pressure and porosity profiles for the Rås and Halten Terrace at 2.5 Ma. Fluid pressures on the Halten Terrace were close to hydrostatic at the onset of the Plio-Pleistocene subsidence, whereas overpressures in the lower Cretaceous–middle Jurassic formations were predicted to be in excess of 40 MPa. The numbers are the lithology codes in Table 2.

mass before pressure build-up and the amount of imported mass:

$$V_{(in)}\rho_{fl(in)} + V_{(before)}\rho_{fl(before)} = V_{(today)}\rho_{fl(today)}. \quad (3)$$

Assuming that the pore volume in the Jurassic reservoirs did not change (as compaction is irreversible) results in $V_{(before)} = V_{(today)} = V$. Then, the additional amount of fluid volume, $V_{(in)}$, which is required to increase the over-pressure in the Jurassic reservoirs on Halten West, is given by

$$V_{(in)} = V(\rho_{fl(today)} - \rho_{fl(before)})/\rho_{fl(in)}. \quad (4)$$

Combining equations (2) and (4) gives

$$V_{(in)} = V(\rho_{fl(before)}\beta\partial P_{fl})/\rho_{fl(in)}. \quad (5)$$

Using equation (5), the volume of fluids required to increase the overpressure on Halten Terrace can be calculated. Assuming a fluid compressibility β of 0.5×10^{-9} Pa^{-1}, a fluid pressure increase of 30 MPa and $\rho_{fl(before)} = \rho_{fl(in)} = 1000$ kg/m^3 gives

$$V_{(in)} = V(1000 \times 0.5 \times 10^{-9} \times 30 \times 10^6)/1000$$
$$= 0.015 \times V.$$

Thus, the volume required to increase the fluid pressure by 30 MPa is c. 1.5% of the pre-existing pore fluid volume on the Halten Terrace, which is only a small fraction. The actual fluid volume that is accessible in the Rås Basin is, however, difficult to assess. The volumes of pore fluids in internal communication include not only the middle Jurassic formations, but also permeable layers (sandstones, siltstones) in the overpressured lower Cretaceous formations. To imagine the effect of this fluid volume withdrawn from the Rås Basin, we have to take into account that the fluids originate from a volume approximately three times larger than that of Halten West. Therefore, the fraction of fluid transported out of the Rås Basin probably was <0.5%. Using similar reasoning as above, but now starting with a known fraction of mass change, the decrease in fluid pressure in the Rås Basin as a result of the fluid loss would be <10 MPa.

Mechanisms for pressure release

In the analysis of porosity–fluid pressure relationships in the middle Jurassic formations, it was shown that porosities on Halten West are only slightly higher than in the low-pressure regime. From the porosity observations we infer that pressure transfer from the Rås Basin occurred during a late stage in the evolution of the margin. If the overpressured fluids migrated from the Rås Basin to the Halten Terrace the most likely migration path is through and/or along the Klakk Fault Complex. The permeability of the Klakk Fault Complex must then have changed from being relatively tight to open, to build up and preserve high overpressure in the Rås Basin during Cretaceous time and to release the overpressure at a later stage (Fig. 16). Furthermore, the fault zone separating the reservoirs on the westernmost part of the Halten Terrace from the Smørbukk reservoirs was apparently closed during and after the episode of pressure transport, as evidenced by the different overpressures of Halten West and the Smørbukk field.

A fault can influence fluid migration in several ways. The fault offset influences fluid flow patterns by juxtaposing units with different hydraulic properties. The damaged rock in the fault zone has hydraulic characteristics that differ from those of the non-damaged host rock. However, the juxtaposition cannot account for sudden changes in fluid flow patterns and overpressure distribution a long time after faulting. The permeability of damaged rocks in the fault zone is partly determined by stresses acting on them. For example, fractures located in the fault zone close when the maximum horizontal stress direction is at a high angle to the fault zone, but open when the minimum horizontal stress is at a high angle to the fault (Grauls & Baleix 1994; Sibson 1994). In addition, opening and closure of fractures in a fault zone during a faulting event results in changed permeability characteristics on a short time-scale (Muir Wood 1994; Sibson 1994). Therefore, the permeability of fault zones is a dynamic factor in a basin's fluid dynamic history, and it is to a large extent stress dependent. Changing basinal stress conditions can, therefore, explain the varying permeability of fault zones with time. In the next section we briefly review the stress history of the western part of the Halten Terrace.

Recent–Neogene intraplate stress history

The Norwegian continental margin is currently characterized by intermediate to low seismicity. Several major fault zones of the mid-Norwegian margin (the Møre–Trøndelag Fault Zone, the Rana Fault Complex and the Kristiansund–Bodø Fault Zone) have been seismically active during the 1800–1989 time period (Gabrielsen 1989; Bungum et al. 1991). Earthquake focal mechanism solutions give NW–SE maximum horizontal stress directions (Bungum et al. 1991; Fejerskov et al. 1995; Lindholm et al. 1995b), which indicates a connection between the intraplate stress field and North Atlantic ridge push. The same stress field extends into the North Sea Basin (Lindholm et al. 1995a), south Sweden–Denmark (Wahlström 1993) and neighbouring areas (Gregersen 1992).

Detailed borehole breakout information is available for seven wells on the Halten Terrace

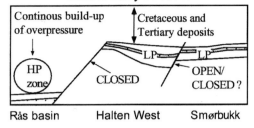

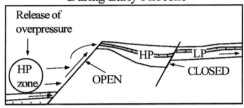

Fig. 16. A schematic model that illustrates how lateral pressure transfer from the Rås Basin and into the Halten Terrace area could possibly explain the present-day fluid pressure distribution in the Jurassic formations. The model assumes that the hydraulic properties of the Klakk Fault Complex changed during an event related to a basinal stress-reorientation in late Neogene (early Pliocene) time. The fault zone separating the overpressured reservoir on Halten West from the close to normally pressured reservoirs of the Smørbukk field must have been closed when the pressure transfer took place, although the exact timing of this closure is not clear. HP and LP denote formations with highly overpressured pore fluids (>25 MPa) and close to hydrostatic fluid pressure (<7.5 MPa), respectively. (See text for discussion.)

(Borgerud & Svare 1995). Additional borehole breakout data have been given by Gölke (1996). In general, the borehole break-out directions confirm the generalized NW–SE orientation of the maximum horizontal stress direction. A compilation of stress-directional data for the mid-Norwegian margin is given in Fig. 17.

A detailed study of palaeo-stress directions on the Halten Terrace is lacking. However, close to the terrace several compressional domes and arches occur (Fig. 17), which provide indirect indications for palaeostress directions. The compressional structures are part of a system of domes and reverse faults (inversion structures) located on the NE Atlantic margin (Doré & Lundin 1996). The domes have two growth phases (Blystad et al. 1995). The oldest phase occurred during Eocene time. The timing of the second phase differs among the domes. The domes located along the southern and western margins of the Halten Terrace (Helland Hansen

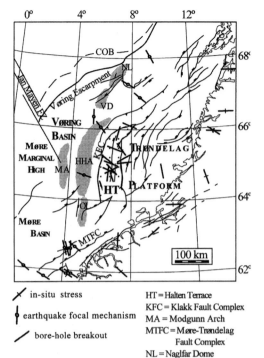

Fig. 17. Stress and domes map for the mid-Norwegian margin. Data from Borgerud & Svare (1995), Fejerskov *et al.* (1995) and Gölke (1996). (See text for discussion.)

Arch, Modgunn Arch, Ormen Lange Dome) have their second growth phase during late Miocene time. The domes in the northern part of the mid-Norwegian margin have a second tectonic phase starting in late Pliocene time which may have continued up to recent times (Blystad *et al.* 1995). The structures were probably formed in response to forces resulting from the farfield effects of Alpine tectonics during an early stage, combined with wrench tectonics along major shear zones that are connected to oceanic transform faults (Doré & Lundin 1996). In addition, an increasing effect of ridge push from the North Atlantic spreading zone was present during the later stage of development.

The southern domes are close to the Klakk Fault Complex, the fault zone separating the western part of the Halten Terrace from the deep and presumably highly overpressured Rås Basin. Therefore, the tectonic history of these arches best documents the palaeostress history of the area of interest. The fact that the second growth phase occurred during late Miocene time implies that a change in the lithospheric stress system of this part of the mid-Norwegian margin occurred immediately after it, during early Pliocene time. The southern domes are N–S oriented, implying an E–W directed maximum compressive stress direction in the area of the domes. During early Pliocene time the stress system may have changed to the present-day situation, i.e. NW–SE directed compression. Similar Neogene stress reorientations have also been documented in areas surrounding the mid-Norwegian margin (e.g. Bergerat 1987; Masson *et al.* 1994; Hibsch *et al.* 1995; Gudmundson *et al.* 1996). The 45° reorientation of the maximum compressive stress direction during early Pliocene time may have caused changes in the permeability of fault and fracture systems by dynamic opening and closure. The Klakk Fault Complex is N–S oriented. Therefore, during late Miocene time it can be inferred that this fault was closed by the E–W maximum horizontal stress direction. The early Pliocene stress reorientation may have resulted in opening of the Klakk Fault Complex along weak spots, which made the pressure transfer from the deep Rås Basin to the western part of the Halten Terrace possible.

The exact timing and processes involved in the closure of the fault separating the highly overpressured formations on Halten West from the normally pressured formations in the Smørbukk area are not known. Olstad *et al.* (1997) suggested that this fault zone became a barrier to fluid flow as a result of diagenetic processes during Plio-Pleistocene time. An alternative (or supplementary) explanation is that the closure is related to secondary stresses resulting from fault–stress interaction at the Klakk Fault Complex, as predicted by the modelling results of Van Balen *et al.* (in prep.). In this model, intraplate stress-induced flexural subsidence at the footwall of the Klakk Fault Complex causes compressional bending stresses occurring at the fault separating the western part of the Halten Terrace from the Smørbukk field. These stresses have an opposite sign to the regional intraplate stress, and in the model they cause closure of the fault.

Discussion and conclusions

In this paper, it has been shown that porosities in the middle Jurassic and apparently also in the lower Cretaceous formations on the western part of the Halten Terrace are only slightly influenced by the present-day high overpressures. The porosity observations are not compatible with local mechanical compaction as the sole cause of the overpressures. Moreover, predictions of undercompaction suggest that overpressuring occurred late, as overpressures inherited from pre-Pliocene times would have resulted in more pronounced undercompaction on the western part of the Halten Terrace. Overpressure development as a result of Plio-Pleistocene loading alone can account only for approximately one-third of the overpressure magnitudes observed in the highly overpressured Jurassic reservoirs.

To explain the observed high fluid pressure–low porosity relationships, we have proposed a model that includes pressure transfer from the nearby deep Rås Basin. This basin is separated from the Halten Terrace by the Klakk Fault Complex. A reorientation of the maximum horizontal stress direction, documented by the growth phases of compressional arches and domes near the fault zone, may have been responsible for increasing the permeability of damaged rocks within the Klakk Fault Complex. The resulting fluid pressure transfer from the deep basin may have caused a sudden fluid pressure increase at the western part of the Halten Terrace, which resulted in local hydraulic fracturing with simultaneous leakage of hydrocarbons and redistribution of overpressure into Cretaceous sandstones. The proposed model implies that hydrocarbon leakage occurred during early Pliocene time, earlier than previously suggested by Ungerer et al. (1990).

Other studies have suggested that fluid transfer may have played a role in the pressure distribution on the Norwegian margin. Koch & Heum (1995) attributed overpressures on Halten West to the rapid subsidence during Plio-Pleistocene time, possibly assisted by lateral transfer of high fluid pressures from the deeply buried Møre Basin to the southwest (located south of the Rås Basin). In the North Sea, Wensaas et al. (1994) argued that high fluid pressures in the Upper Cretaceous Shetland Group in the Gullfaks Field are partly due to pressure transport from a deeper part of the basin, as overpressure generating mechanisms such as compaction disequilibrium, hydrocarbon generation, clay dehydration and aquathermal expansion failed to explain these high fluid pressures. Furthermore, Caillet et al. (1991) suggested that the high fluid pressures in the Snorre Field, northern North Sea, were originally generated in the deep parts of the basin. These workers suggested that these pressures were transmitted to the Snorre reservoirs through direct communication or through fractures or other discontinuities in the shale formations. Thus, fluid (pressure) transfer appears to have played an important role in the North Sea and must be considered as a potential process that may have affected the fluid pressure distribution on the Halten Terrace.

Although it is not the primary objective of this paper to thoroughly evaluate overpressure mechanisms besides compaction disequilibrium (local and external), some discussion about other potential explanations for the high fluid pressure–low porosity relationships is appropriate. Our observations could also be explained by sediment unloading (late stage overpressuring related to Plio-Pleistocene rapid subsidence), or shale compaction at greater depths may occur without hindrance by fluid overpressures (as proposed by Hermanrud et al. (1997)). Possible mechanisms for late-stage fluid pressure generation include gas generation, mineralogical transformations, aquathermal expansion and lateral compression. The effect of aquathermal expansion has by many workers been considered to be much less important than compaction disequilibrium in developing overpressures (Bethke 1986; Shi & Wang 1986). Luo & Vasseur (1992) and Osborne & Swarbrick (1997) concluded that this process is insignificant in most geological situations. Tectonic compression (horizontal loading) as a cause of overpressuring has been reported from tectonically active basins (e.g. Berry 1973). The mid-Norwegian margin is currently in a compressive state of stress and, therefore, lateral compression could be a factor in overpressuring. However, the effects of lateral stresses on overpressuring are not well understood, mainly because of the geological complexities of basins in which lateral compression occurs (Osborne & Swarbrick 1997).

Gas generation or cracking of oil to gas is accompanied by a large volume expansion (Ungerer et al. 1983; Barker 1990) and has therefore the potential to be a major factor in overpressuring. The main source rocks on the Halten Terrace are the lower Jurassic gas-prone Åre Formation and the upper Jurassic oil-prone Spekk Formation. Maturity of the Åre Formation is greatest in the western part of the area, where the formation is in the wet to dry gas window (Vik et al. 1992). Pressure communication with gas mature shales from deeper parts on Halten West could possibly explain the high overpressures in the Jurassic reservoirs. However, most of the highly overpressured Jurassic reservoirs are today without accumulations of hydrocarbons, which suggests that present-day *in situ* gas generation is not a primary cause of the high overpressures. However, such an explanation cannot be ruled out. Hydrocarbon generation as the cause of the overpressures in the Cretaceous formations seems less likely, as there are no well-defined source rocks within the post-Jurassic successions.

The role of diagenetic processes as a mechanism to reduce the porosity and increase the fluid pressure has been emphasized by Bjørlykke & Høeg (1997). Below 2–3 km, important mineralogical changes take place in shales, which are mainly functions of temperature and mineralogy rather than effective stress (Bjørlykke & Høeg 1997). The effect of these mineralogical transformations (e.g. smectite dehydration and smectite–illite transformation) for increasing the fluid pressure is under debate, mainly because of the uncertainties related to whether sufficient volumes of fluids are released during the transformations. Osborne & Swarbrick (1997) concluded that smectite dehydration is unlikely to be a primary cause of overpressuring in sedimentary basins because the volume of fluids released is small. Audet (1995) modelled the effect of mechanical compaction and clay dehydration on overpressure development and showed that clay dehydration can potentially increase overpressure by about 30%. The potential for mineral transformations to have produced parts of the very high fluid pressures on Halten Terrace must be considered. However, a better knowledge of the sediment composition as well as the reactions that may have taken place is required.

Hermanrud et al. (1998) suggested that compaction could possibly continue without hindrance by overpressure, on the basis of their study of the porosity distribution in overpressured and normally pressured shales on the Norwegian margin. Their main argument was that it is unlikely that the overpressures in the close to normally compacted Jurassic shales on the Halten Terrace and in the North Sea Cretaceous shales are all due to late-stage sediment unloading. The observations from the Cretaceous shales on the Halten Terrace (Fig. 10) suggest that these shales are close to normally compacted. However, the generally poor control on the fluid pressure distribution in these shales combined with the lack of detailed information about normal compaction trends makes it difficult to assess whether the shales are, in fact, normally compacted and highly overpressured. An interpretation of the porosity–fluid pressure relationships in the Cretaceous formations is that the shales are slightly undercompacted with respect to the Plio-Pleistocene sediment loading (which would be expected, because of the high sedimentation rates during Plio-Pleistocene time). The high fluid pressures in the sandstones may be due to local compaction disequilibrium assisted by vertical (or lateral) pressure transfer from the deeper highly overpressured reservoirs.

In summary, the high overpressures and low porosities in the middle Jurassic formations on the western part of the Halten Terrace may have been produced by a combination of several processes, although the relative importance of some of these processes is difficult to assess. In this paper, we have presented a model for the overpressure evolution that is compatible with the observations of the porosity distribution. The present study has illustrated how porosity–fluid pressure observations can be explained by linking the temporal changes in basinal stress systems to changes in fault properties, and that these interactions may have had major implications for fluid flow systems on the Halten Terrace. More work is needed to further confirm this hypothesis. A better understanding of these transfer phenomena will have direct implications for predictions of anomalous pressures, the efficiency of caprocks and for secondary migration of hydrocarbons.

This research was funded through Norsk Hydro and is part of a continuing project between Norsk Hydro and Vrije Universiteit. We thank D. Hermansen from Norsk Hydro for stimulating discussions and provision of data during the progress of this work. B. T. Larsen from Saga Petroleum is thanked for his comments on an earlier version of this manuscript. H. Kooi is appreciated for helpful comments on the modelling part. The paper benefited from the helpful reviews of C. Hermanrud and M. Osborne. Norsk Hydro is gratefully acknowledged for giving permission to publish. This is Netherlands Research School of Sedimentary Geology Publication 990206.

References

APLIN, A. C., YANG, Y. & HANSEN, S. 1995. Assessment of β, the compression coefficient of mudstones and its relationship with detailed lithology. *Marine and Petroleum Geology*, **12**, 955–963.

AUDET, D. M. 1995. Mathematical modelling of gravitational compaction and clay dehydration in thick sediment layers. *Geophysical Journal International*, **122**, 283–298.

—— & MCCONNELL, J. D. C. 1992. Forward modelling of porosity and pore pressure evolution in sedimentary basins. *Basin Research*, **4**, 147–162.

BARKER, C. 1990. Calculated volume and pressure changes during the thermal cracking of oil to gas in reservoirs. *Bulletin, American Association of Petroleum Geologists*, **74**, 1254–1261.

BERGERAT, F. 1987. Stress fields in the European platform at the time of Africa–Eurasia collision. *Tectonics*, **6**, 99–132.

BERRY, F. 1973. High fluid potentials in California Coast Ranges and their tectonic significance. *Bulletin, American Association of Petroleum Geologists*, **57**, 1219–1249.

BETHKE, C. M. 1986. Inverse hydrologic analysis of the distribution and origin of Gulf Coast-type geopressured zones. *Journal of Geophysical Research*, **91**, 6535–6545.

BJØRLYKKE, K. & EGEBERG, P. K. 1993. Quartz cementation in sedimentary basins. *Bulletin, American Association of Petroleum Geologists*, **77**, 1538–1548.

—— & HØEG, K. 1997. Effects of burial diagenesis on stresses, compaction and fluid flow in sedimentary basins. *Marine and Petroleum Geology*, **14**, 267–276.

BLYSTAD, P., BRE .KE, H., FÆRSETH, R. B., LARSEN, B. T., SKOGSEID, J. & TØRUDBAKKEN, B. 1995. *Structural elements of the Norwegian continental shelf, part II: The Norwegian Sea Region.* Norwegian Petroleum Directorate Bulletin, **8**.

BORGERUD, L. K. & SVARE, E. 1995. In-situ stress field on the Norwegian Margin, 62°–67° North. *In*: FEJERSKOV, M. & MYRVANG, A. M. (eds) *Proceedings of the Workshop Rock Stresses in the North Sea, Trondheim, Norway*, 165–178.

BREDEHOEFT, J. D., DJEVANSHIR, R. D. & BELITZ, K. R. 1988. Lateral fluid flow in a compacting sand–shale sequence: South Caspian Sea. *Bulletin, American Association of Petroleum Geologists*, **72**, 416–424.

BUNGUM, H., ALSAKER, A., KVAMME, L. B. & HANSEN, R. A. 1991. Seismicity and seismotectonics of Norway and nearby continental shelf areas. *Journal of Geophysical Research*, **96**, 2249–2265.

CAILLET, G., SÉJOURNÉ, C., GRAULS, D. & ARNOUD, J. 1991. The hydrodynamics of the Snorre Field area, offshore Norway. *Terra Nova*, **3**, 180–194.

DALLAND, A., WORSLEY, D. & OFSTAD, K. 1988. *A lithostratigraphic scheme for the Mesozoic and Cenozoic succession offshore mid- and northern Norway.* Norwegian Petroleum Directorate Bulletin, **4**.

DICKINSON, G. 1953. Geological aspects of abnormal reservoir pressure in Gulf Coast Louisiania. *Bulletin, American Association of Petroleum Geologists*, **37**, 410–432.

DORÉ, A. G. & LUNDIN, E. R. 1996. Cenozoic compressional structures on the NE Atlantic margin: nature, origin and potential significance for hydrocarbon exploration. *Petroleum Geoscience*, **2**, 299–311.

EHRENBERG, S. N. 1990. Relationship between diagenesis and reservoir quality in sandstones of the Garn Formation, Haltenbanken, mid-Norwegian continental shelf. *Bulletin, American Association of Petroleum Geologists*, **74**, 1538–1558.

——, GJERSTAD, H. M. & HADLER-JACOBSEN, F. 1992. Smørbukk Field — a gas condensate fault trap in the Haltenbanken province, offshore mid-Norway. *In*: HALBOUTY, M. T. (ed.) *Giant Oil and Gas Fields of the Decade 1978–1988.* American Association of Petroleum Geologists Memoir, **54**, 323–348.

FEJERSKOV, M., MYRVANG, A. M., LINDHOLM, C. & BUNGUM, H. 1995. In-situ rock stress pattern on the Norwegian continental shelf and mainland. *In*: FEJERSKOV, M. & MYRVANG, A. M. (eds) *Proceedings of the Workshop Rock Stresses in the North Sea, Trondheim, Norway*, 191–201.

FERTL, W. H., CHAPMAN, R. E. & HOTZ, R. F. 1994. *Studies in Abnormal Pressures.* Elsevier, New York.

GAARENSTROOM, L., TROMP, R. A. J., DE JONG, M. C. & BRANDENBURG, A. M. 1993. Overpressures in the Central North Sea: implications for trap integrity and drilling safety. *In*: PARKER, J. R. (ed.) *Petroleum of Northwest Europe: Proceedings of the 4th Conference.* Geological Society, London, 1305–1313.

GABRIELSEN, R. H. 1989. Reactivation of faults on the Norwegian continental shelf and its implications for earthquake occurrence. *In*: GREGERSEN, S. & BASHAM, P. W. (eds) *Earthquakes of North-Atlantic Passive Margins: Neotectonics and Postglacial Rebound, NATO ASI Series.* Kluwer, Dordrecht, 67–90.

GÖLKE, M. 1996. *Patterns of stress in sedimentary basins and the dynamics of pull-apart basin formation.* PhD thesis, Vrije Universiteit, Amsterdam.

GRAULS, D. J. & BALEIX, J. M. 1994. Role of overpressures and *in situ* stresses in fault-controlled hydrocarbon migration: a case study. *Marine and Petroleum Geology*, **11**, 734–742.

GREEN, D. H. & WANG, H. F. 1986. Fluid pressure response to undrained compression in saturated sedimentary rock. *Geophysics*, **51**, 948–956.

GREGERSEN, S. 1992. Crustal stress regime in Fennoscandia from focal mechanisms. *Journal of Geophysical Research*, **97**, 11821–11827.

GUDMUNDSSON, A., BERGERAT, F. & ANGELIER, J. 1996. Off-rift and rift-zone palaeostresses in Northwest Iceland. *Tectonophysics*, **255**, 211–228.

HAGEVANG, T. & TØRUDBAKKEN, B. 1996. På dypt vann i Norskehavet. *Geonytt*, **2**, 21–26.

HANSEN, S. 1996. A compaction trend for Cretaceous and Tertiary shales on the Norwegian Shelf based on sonic transit times. *Petroleum Geoscience*, **2**, 159–166.

HERMANRUD, C., WENSAAS, L., TEIGE, G. M. G., VIK, E., BOLAAS, H. M. N. & HANSEN, S. 1998. Shale porosities from well logs on Haltenbanken (offshore Mid-Norway) show no influence of overpressuring. *In*: LAW, B. (ed.) *Abnormal Pressures in Hydrocarbon Environments.* American Association of Petroleum Geologists Memoir, **70**, 65–85.

HIBSCH, C., JARRIGE, J. J., CUSHING, E. M. & MERCIER, J. 1995. Palaeostress analysis, a contribution to the understanding of basin tectonics and geodynamic evolution. Example of the Permian/Cenozoic tectonics of Great Britain and geodynamic implications in western Europe. *Tectonophysics*, **252**, 103–136.

KOCH, J. O. & HEUM, O. R. 1995. Exploration trends of the Halten Terrace. *In*: HANSLIEN, S. (ed.), *Petroleum Exploration and Exploitation in Norway.* Norwegian Petroleum Society (NPF), Special Publication, **4**, 231–251.

KOOI, H. 1997. Insufficiency of compaction disequilibrium as the sole cause of high pore fluid pressures in pre-Cenozoic sediments. *Basin Research*, **9**, 227–241.

LINDHOLM, C. D., BUNGUM, H., BRATLI, R. K., AADNØY, B. S., DAHL, N. TØRUDBAKKEN, B. & ATAKAN, K. 1995a. Crustal stress in the northern North Sea as inferred from borehole breakouts and earthquake focal mechanisms. *Terra Nova*, 7, 51–59.

——, ——, VILLAGRAN, M. & HICKS, E. 1995b. Crustal stress and tectonics in Norwegian regions determined from earthquake focal mechanisms. *In*: FEJERSKOV, M. & MYRVANG, A. M. (eds) *Proceedings of the Workshop Rock Stresses in the North Sea, Trondheim, Norway*, 77–91.

LUO, X. & VASSEUR, G. 1992. Contributions of compaction and aquathermal pressuring to geopressure and the influence of environmental conditions. *Bulletin, American Association of Petroleum Geologists*, 76, 1550–1559.

—— & —— 1993. Compaction coefficients of argillaceous sediments: their implications, significance and determination. *In*: DORÉ, A. G., AUGUSTSON, J. H., HERMANRUD, C., STEWART, D. J. & SYLTA, Ø. (eds) *Basin Modelling: Advances and Applications*. Norwegian Petroleum Society (NPF), Special Publication, 3, 201–217.

MAGARA, K. 1976. Water expulsion from clastic sediments during compaction — directions and volumes. *Bulletin, American Association of Petroleum Geologists*, 60, 543–553.

MANN, D. M. & MACKENZIE, A. S. 1990. Prediction of pore fluids in sedimentary basins. *Marine and Petroleum Geology*, 7, 55–65.

MASSON, D. G., CARTWRIGHT, J. A., PINHEIRO, L. M., WHITMARSH, R. B., BESLIER, M. O. & ROESER, H. 1994. Compressional deformation at the ocean–continent transition in the NE Atlantic. *Journal of the Geological Society, London*, 151, 607–613.

MUIR WOOD, R. 1994. Earthquakes, strain-cycling and the mobilization of fluids. *In*: PARNELL, J. (ed.) *Geofluids: Origin, Migration and Evolution of Fluids in Sedimentary Basins*. Geological Society, London, Special Publications, 78, 85–98.

OLSTAD, R., BJØRLYKKE, K. & KARLSEN, D. A. 1997. Pore water flow and petroleum migration in the Smørbukk field area, offshore mid-Norway. *In*: MØLLER-PEDERSEN, P. & KOESTLER, A. G. (eds) *Hydrocarbons Seals: Importance for Exploration and Production*. Norwegian Petroleum Society (NPF), Special Publication, 7, 201–217.

OSBORNE, M. J. & SWARBRICK, R. E. 1997. Mechanisms for generating overpressure in sedimentary basins: a reevaluation. *Bulletin, American Association of Petroleum Geologists*, 81, 1023–1041.

PALCIAUSKAS, V. V. & DOMENICO, P. A. 1982. Characterization of drained and undrained response of thermally loaded repository rocks. *Water Resources Research*, 18, 281–290.

RAMM, M. & BJØRLYKKE, K. 1994. Porosity/depth trends in reservoir sandstones: assessing the quantitative effects of varying pore-pressure, temperature history and mineralogy, Norwegian shelf data. *Clay Minerals*, 29, 475–490.

RIIS, F. 1996. Quantification of Cenozoic vertical movements of Scandinavia by correlation of morphological surfaces with offshore data. *Global and Planetary Change*, 12, 331–357.

SHI, Y. & WANG, C.-Y. 1986. Pore pressure generation in sedimentary basins: overloading versus aquathermal. *Journal of Geophysical Research*, 91, 2153–2162.

SIBSON, R. H. 1994. Crustal stress, faulting and fluid flow. *In*: PARNELL, J. (ed.) *Geofluids: Origin, Migration and Evolution of Fluid in Sedimentary Basins*. Geological Society, London, Special Publications, 78, 69–84.

SKOGSEID, J., PEDERSEN, T. & LAFSEN, V. B. 1992. Vøring Basin: subsidence and tectonic evolution. *In*: LARSEN, R. M., BREKKE, H., LARSEN, B. T. & TALLERAAS, E. (eds) *Structural and Tectonic Modelling and its Application to Petroleum Geology*. Norwegian Petroleum Society (NPF), Special Publication, 1, 55–82.

UNGERER, P., BEHAR, E. & DISCAMPS, D. 1983. Tentative calculation of the overall volume expansion of organic matter during hydrocarbon genesis from geochemistry data. Implications for primary migration. *In*: BJORØY, M. (ed.) *Advances in Organic Geochemistry 1981*: Wiley, Chichester, 129–135.

——, BURRUS, J., DOLIGEZ, B., CHÉNET, P. Y. & BESSIS, F. 1990. Basin evaluation by integrated two-dimensional modelling of heat transfer, fluid flow, hydrocarbon generation, and migration. *Bulletin, American Association of Petroleum Geologists*, 74, 309–335.

VERNIK, L. 1997. Predicting porosity from acoustic velocities in siliciclastics: a new look. *Geophysics*, 62, 118–128.

——, L. & NUR, A. 1992. Petrophysical classification of siliciclastics for lithology and porosity prediction from seismic velocities. *Bulletin, American Association of Petroleum Geologists*, 76, 1295–1309.

VIK, E., HEUM, O. R. & AMALIKSEN, K. G. 1992. Leakage from deep reservoirs: possible mechanisms and relationship to shallow gas in the Haltenbanken area, mid-Norwegian Shelf. *In*: ENGLAND, W. A. & FLEET, A. J. (eds) *Petroleum Migration*. Geological Society, London, Special Publications, 59, 273.

WAHLSTRØM, R. 1993. Fennoscandian seismicity and its relation to the isostatic rebound. *Global and Planetary Change*, 8, 107–112.

WARD, C. 1995. Evidence for sediment unloading caused by fluid expansion overpressure-generating mechanisms. *In*: FEJERSKOV, M. & MYRVANG, A. M. (eds) *Proceedings of the Workshop Rock Stresses in the North Sea, Trondheim, Norway*, 218–231.

WENSAAS, L., SHAW, H. F., GIBBONS, K., AAGAARD, P. & DYPVIK, H. 1994. Nature and causes of overpressuring in mudrocks of the Gullfaks area, North Sea. *Clay Minerals*, 29, 439–449.

Porosity characteristics of Cambrian mudrocks (Oak Ridge, East Tennessee, USA) and their implications for contaminant transport

J. DORSCH[1,2] & T. J. KATSUBE[3]

[1]*Environmental Sciences Division, Oak Ridge National Laboratory, Oak Ridge, TN 37831-6400, USA and Department of Geological Sciences, University of Tennessee, Knoxville, TN 37996, USA*
[2]*Present address: Department of Earth and Atmospheric Sciences, St Louis University, St Louis, MO 63103, USA*
[3]*Geological Survey of Canada, Mineral Resources Division, Ottawa, Ont. K1A 0E8, Canada*

Abstract: Thirty-eight mudrock samples from the Cambrian Conasauga Group were analysed by immersion, helium and mercury porosimetry. The data generated from these techniques include effective porosities, pore throat size distributions and sample densities. The purpose of these analyses was to obtain information required to evaluate the capacity of the Conasauga Group mudrock to retard transport of and to store contaminants on the Oak Ridge Reservation (East Tennessee, USA). Results indicate that the three porosities were $9.90 \pm 2.61\%$, $3.8 \pm 0.7\%$ and $8.1 \pm 4.3\%$ for immersion (ϕ_I), mercury (ϕ_{Hg}) and helium porosity (ϕ_{he}), respectively. These values are considerably higher than those previously reported for these rocks (0.1–3.4%). The ϕ_{Hg} values are commonly the smallest of the three. Although the three porosities generally show little change with depth, an expected trend as a result of the maximum burial depth of the Conasauga Group of >4 km, a decrease with depth is seen in some of the coreholes for only ϕ_I. Furthermore, inconsistencies between the different porosity types provide an insight into the mudrock pore structure and related geochemical characteristics. For example, they indicate the existence of considerable pore space accessed only through <3 nm pore throats and the possibility of meteoric cement having replaced original diagenetic cement for some mudrock samples, which would lead to a possible instability to immersion with contaminated water. These characteristics have significant implications for the evaluation of the mudrock retardation and storage capacity for contaminants, and the design of remediation measures.

Thirty-eight mudrock samples from the Conasauga Group (Middle and Upper Cambrian) were analysed by immersion, helium and mercury porosimetry. The purpose of this study was to obtain information on the porosity characteristics of this Cambrian mudrock. This information is required to evaluate the capacity of the Conasauga Group mudrock to retard the transport of and to store contaminants from facilities on the Oak Ridge Reservation (East Tennessee, USA). Although some petrophysical information exists on younger mudrock from Jurassic and Tertiary formations (e.g. Katsube & Williamson (1994) and references therein), little is known to exist on Middle and Upper Cambrian mudrocks. Moreover, these data will provide constraints for modelling hydrological behaviours in the mudrock-dominated Conasauga Group, the most significant aquitard on the Oak Ridge Reservation.

It has been reported that effective porosity is one of the most important factors controlling diffusive processes within fine-grained siliciclastic rocks (e.g. Germain & Frind 1989; Toran *et al.* 1995). In this paper, effective porosity represents all interconnected pore space; this implies all pores such as connecting, storage and dead-end pores (Katsube *et al.* 1994). Matrix diffusion is considered to be an important mechanism related to transport of material in the fractured low-permeability sedimentary rocks at the Oak Ridge Reservation (Wickliff *et al.* 1991; Solomon *et al.* 1992; Sandford *et al.* 1994; Shevenell *et al.* 1994). This causes contaminants in water moving through the interconnected fracture network to have access to the interconnected pore space of

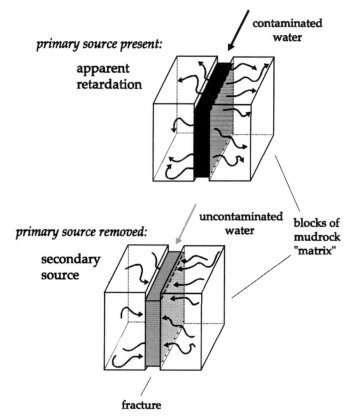

Fig. 1. Block diagrams illustrating diffusion of contaminants from a fracture into the mudrock 'matrix' (apparent retardation) and vice versa (secondary contaminant source).

the matrix blocks by diffusion (Fig. 1), which results in the apparent retardation of the spreading of contaminant species (e.g. Neretnieks 1980; Tang et al. 1981; McKay et al. 1993). Subsequent to removal of the primary contaminant source, by either remediation or depletion of the source, the contaminants will diffuse back into the fracture network (Fig. 1) and create a secondary contaminant source (e.g. Germain & Frind 1989; McKay et al. 1993). This secondary process could be active for a long period and may necessitate costly additional remediation efforts. Accurate knowledge of the porosity characteristics is, therefore, essential for evaluating possible retardation of contaminant spread and development of secondary contaminant sources within mudrock (Toran et al. 1995), although the characteristics of fluids associated with the rock will also influence diffusive processes within the rock matrix.

The Oak Ridge Reservation is situated within the Appalachian Valley-and-Ridge physiographic province in East Tennessee. This physiographic province corresponds to the Alleghanian foreland fold-and-thrust belt. Stratigraphic units contained within different thrust sheets range in age from Early Cambrian to Early Pennsylvanian and are repeated and stacked (Fig. 2; Hatcher et al. 1992), because the original fill of the Appalachian basin was deformed during thrusting associated with the Alleghanian orogeny. The stratigraphic units on the Oak Ridge Reservation have been grouped into regional aquifers and aquitards (Fig. 3; Solomon et al. 1992). This study focuses on the mudrock-dominated Conasauga Group (Middle and Upper Cambrian), because the majority of waste sites on the Oak Ridge Reservation are located on Conasauga Group rocks. The Conasauga Group mudrock appears within the Copper Creek thrust sheet and the Whiteoak Mountain thrust sheet (Fig. 4). Geographically, the stratigraphic units underlie Melton Valley in the Copper Creek thrust sheet and Bear Creek Valley in the Whiteoak Mountain thrust sheet. The Conasauga Group represents the fill of a

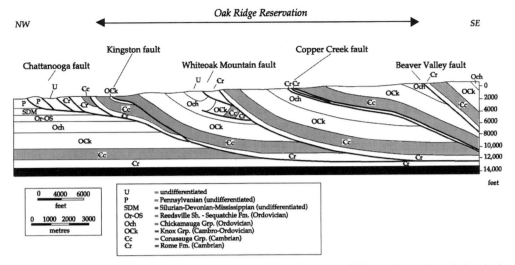

Fig. 2. Generalized geological cross-section through part of the Alleghanian fold-and-thrust belt at the longitude of the Oak Ridge Reservation. The stippled bands outline the Conasauga Group stratigraphic units, black outlines Precambrian basement (modified from Hatcher et al. 1992).

Mid- to Late Cambrian intrashelf basin that developed on the Laurentian passive continental margin bordering the evolving Iapetus ocean (e.g. Markello & Read 1981, 1982; Read 1989; Srinivasan & Walker 1993). The fill constitutes an alternation of mudrock- and carbonate-dominated stratigraphic units (Rodgers 1953; Hasson & Haase 1988; Walker et al. 1990). In the area of the Oak Ridge Reservation, the Conasauga Group is dominated by mudrock (Hatcher et al. 1992), in contrast to well-studied areas to the southeast (e.g. Srinivasan & Walker 1993), where the Conasauga Group is dominated by carbonate rocks. Compaction of freshly deposited mud and mudrock is the primary determinant in reduction of original sedimentary porosity. Data on the maximum burial of Conasauga Group mudrock of the Whiteoak Mountain thrust sheet in East Tennessee were provided by Foreman (1991). According to basin-subsidence modelling the Conasauga Group mudrock experienced burial of up to 4 km. Structural evidence, however, indicates that the Copper Creek thrust sheet was on top of the Whiteoak Mountain thrust sheet following the Alleghanian orogeny (Lemiszki, pers. comm.), thus making a greater burial depth more likely. Conasauga Group strata in Virginia, for instance, are believed to have been buried to between 6 and 8.5 km (Mussman et al. 1988; Montanez 1994). A burial depth of 6–7 km might be a better estimate for the Conasauga Group strata on the Oak Ridge Reservation.

Method of investigation

Samples

Forty-two mudrock core-samples were obtained from coreholes drilled into the Whiteoak Mountain (Cores GW-132, GW-133 and GW-134) and Copper Creek (Cores Wol-1 and 0.5MW012A) thrust sheet (Fig. 4). The cores were inspected visually and sampling intervals were selected (Dorsch et al. 1996) on the basis of (a) macroscopic homogeneity of the sampling interval, (b) absence of excessive deformation, (c) mudrock lithology, and (d) stratigraphic coverage. Thirty-eight samples, representing each of the mudrock-dominated stratigraphic units of the Conasauga Group (i.e. Nolichucky Shale, Maryville Limestone, Rogersville Shale, Rutledge Limestone and Pumpkin Valley Shale), were selected for porosity measurements.

The core samples were, commonly, just under 30 cm in length. Irregularly shaped specimens of mudrock were broken off each of the core samples and prepared for one or all of the three types of porosity (immersion, mercury and helium porosimetry) measurements. In addition, five cylindrically shaped specimens, of 2.54 cm (1 inch) in diameter and commonly ≤0.75 cm in thickness, were prepared for immersion porosimetry. The specimens were generally <10 g in weight, and were taken from as close as possible to the centre of core. This was planned to avoid marginal areas of the cores, which might have been subject to invading drilling fluids and clay

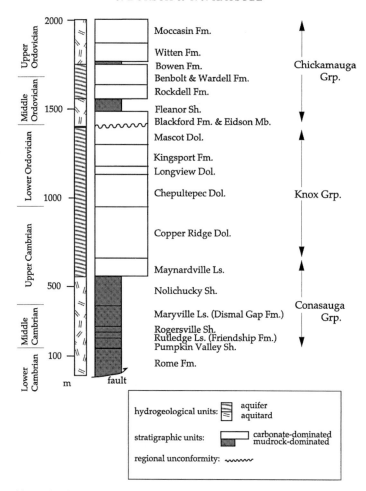

Fig. 3. Stratigraphic section for the Copper Creek and Whiteoak Mountain thrust sheets on the Oak Ridge Reservation, displaying the stratigraphic distribution of mudrock-dominated stratigraphic units (based on Hatcher et al. 1992).

detritus, or coring-induced microfracturing. In addition, macroscopically visible mineralized fractures were also avoided. A total of 120 specimens, from 37 samples, were prepared for immersion porosimetry (11 samples were represented by a multiple number of specimens). Thirty-three specimens, representing the same number of samples, were prepared for both mercury and helium porosimetry.

Porosity measurements, and sample and specimen preparation

Effective porosities for the Conasauga Group mudrock from the Oak Ridge Reservation were measured by three well-established methods: immersion, mercury and helium porosimetry. These methods have all been commonly applied to fine-grained, low-permeability rocks, and have proven to provide useful petrophysical information on tight mudrock (Soeder 1988; Katsube & Scromeda 1991; Katsube & Best 1992; Katsube et al. 1992a; Loman et al. 1993; Issler & Katsube 1994). All specimens for the three porosimetry methods were dried at 100–105°C to ensure measurement of all interconnected pore space available for water storage or intrusion. Drying at these temperatures will drive off all free and adsorbed water (Scromeda & Katsube 1993), but will not affect the crystal-lattice water of the clay minerals (Dorsch 1995). Porosity increases as a result of oven drying are considered to be small. A case has been reported showing the increase to

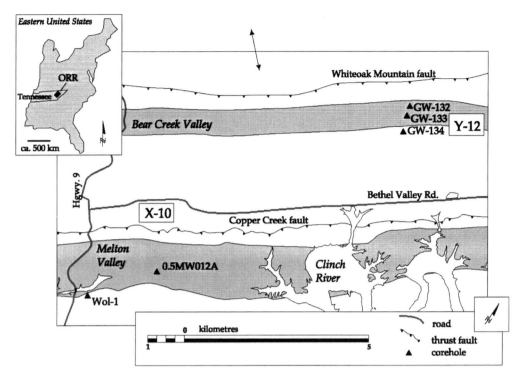

Fig. 4. Generalized location map showing the outcrop of the mudrock-dominated part of the Conasauga Group (dark stippled bands) on the Oak Ridge Reservation (ORR), East Tennessee, and the coreholes from which samples were retrieved. X-10 is the location of the Oak Ridge National Laboratory. Double arrow indicates the approximate orientation of the cross-section (Fig. 2) traversing the Oak Ridge Reservation (based on Hatcher et al. 1992).

be 0–20% of the shale porosity values (Katsube et al. 1992a), which is generally within the ranges of procedural error. Increases in porosity as a result of destressing during sampling are also judged to be small. Porosities resulting from destressing of cored mudrock from depths of 4.0–6.0 km were reported to be in the range of 0.1–0.2% (Katsube et al. 1991). This range is larger than the one reported for crystalline rocks (0–0.15%; Katsube & Marschal 1993), and is similar to values for limestone and dolostone (0–0.22%; Nur & Simmons 1969).

Immersion porosimetry. For the immersion porosimetry method, effective porosity is determined by taking the difference in weight between the fully saturated and dry specimen:

$$\phi_I = \delta_r[(W_W - W_d)/(W_d \delta_w)] \qquad (1)$$

where ϕ_I is the effective porosity determined by immersion porosimetry, δ_r and δ_w are the bulk densities of rock specimen and of water, respectively, and W_W and W_d are the wet and dry weights of the specimen. Saturation of a specimen with water (deionized water) is assumed to penetrate all of the interconnected pore space (Katsube 1992; Katsube et al. 1992a). The immersion porosity (ϕ_I) can also be determined by

$$\phi_I = [(W_W - W_d)/\delta_w]/V_b] \qquad (2)$$

where V_b is the bulk volume of the disc-shaped specimen.

The procedural steps employed at the Geological Survey of Canada (Katsube & Scromeda 1991; Katsube et al. 1992b; Scromeda & Katsube 1993, 1994), and summarized by Dorsch et al. (1996), were followed in this study. The measurement procedure covers 3 days and involves: (a) vacuum degassing of specimen in a vacuum chamber (for 15 min, at an applied vacuum of 27–28.5 inches Hg); (b) submergence of specimen by filling beaker with deionized water; (c) vacuum degassing of submerged specimen in a vacuum chamber (for 15 min, at an applied vacuum of 27–28.5 inches Hg); (d) submerged specimen is left under atmospheric pressure (for 24 h); (e) cautious removal of specimen from beaker, inspection of specimen, careful removal of surface water without drying of specimen surface, and weighing of water-saturated specimen (to provide saturated specimen weight); (f) oven drying of specimen (at 112–116°C, for 24 h); (g) cooling of specimen in desiccator (for 20 min); (h) quick weighing and careful inspection of specimen (to provide dry specimen weight). Specimens used for immersion porosimetry were of irregular or disc shape, as previously indicated. For the irregularly shaped specimens, the ϕ_I values were determined by use of bulk-density values obtained either by mercury porosimetry or by assumption (for some cases), which were inserted into equation (1). For the disc-shaped specimens, the ϕ_I values were determined by measuring the specimen dimensions with a caliper, then calculating the bulk volume and inserting the results into equation (2).

The ϕ_I values were considered reliable only if the specimen survived the experimental procedures completely intact. They were also considered to be reliable if cracking occurred only after the final drying process, because no material would have been lost to cause a change to the dry weight. Further details have been given by Dorsch et al. (1996).

Mercury porosimetry

Mercury porosimetry involves the injection of a non-wetting liquid (mercury) into a specimen in discrete pressure steps using a mercury porosimeter. The pressures required to force mercury into the specimen are related to the sizes of the pore throats by the Washburn Equation (Washburn 1921; Rootare 1970; Wardlaw 1976; Kopaska-Merkel 1988; Wardlaw et al. 1988; Kopaska-Merkel 1991):

$$d = -4\gamma \cos(\theta)/P \quad (3)$$

where d is the pore-throat size, γ is the interfacial (surface) tension (for Hg–vacuum this is 0.48 N m^{-1}), θ is the contact angle (for Hg–vacuum this is 140°) and P is the intrusion pressure (MPa). With each increasing pressure step, successively smaller pore throats are accessed by mercury. The results are initially displayed as capillary-pressure curves of intruded mercury against injection pressure. These are subsequently converted to porosity (partial porosity) v. pore-size data.

Specimens used in this measurement were of irregular shape and generally 2–6 g in weight. The specimens were placed in a Micromeritics Autopore 9200 porosimeter with a pressure capacity of 0.14–420 MPa, which corresponds to an equivalent pore throat size range of about 250 mm–3 nm. The mercury-injection pressure was increased successively in discrete steps (50–60 steps from 0.14 to 420 MPa), more or less equally dividing on a logarithmic scale over the entire pressure range (Katsube & Issler 1993), and the volume of intruded mercury was measured at each step. Equilibrium times were 40 s for the higher-pressure steps (>0.7 MPa) and 10 s for the lower-pressure steps (<0.7 MPa). Accuracy of volume measurement for the intruded mercury is reported to be $< \pm 0.0015$ cm^3 (Kopaska-Merkel 1991).

The volume data are grouped into different pore-size classes, as in previous studies (e.g. Agterberg et al. 1984; Katsube & Walsh 1987; Katsube & Best 1992). The standard display format has each decade of the logarithmic pore-size scale (x-axis) subdivided into five ranges with equal physical spacing. Partial porosity (ϕ_a) represents the porosity calculated for each pore-size range by using the volume of the intruded mercury at this size range and the bulk volume of the specimen. The bulk-specimen volume was determined with calculations involving the penetrometer volume and its weight (with mercury, with specimen and mercury).

The sum of partial porosities (ϕ_a) for pore-size ranges of 3 nm–10 μm and 3nm–250 μm are obtained to generate two effective porosity values (Katsube & Issler 1993): ϕ_{Hg1} and ϕ_{Hg2}. The reason for generating the two values is that ϕ_{Hg2} is likely to contain measurement errors caused by space left between the specimen and its container wall. The ϕ_{Hg1} is more likely to reflect a true effective porosity value free of error (Katsube & Issler 1993). Further details of the procedures have been given by Dorsch et al. (1996). Effective porosity determined by this method is referred to as mercury porosity (ϕ_{Hg}) in this study, and is represented by the ϕ_{Hg1} values.

Helium porosimetry. For helium porosimetry, effective porosity is determined by first placing the specimen into a steel chamber (known volume) of the Boyle's law double-celled helium porosimeter (Loman et al. 1993). Helium is then allowed to isothermally expand into the chamber from a reservoir of known volume and pressure until equilibrium pressure is reached (20–60 min). The grain volume can be calculated from measurement of the new gas pressure. The bulk volume of the specimen is determined by immersion of the specimen in mercury.

Effective porosity is calculated by taking the difference between the grain volume and bulk volume:

$$\phi_{he} = (V_b - V_g)/V_b \qquad (4)$$

where ϕ_{he} is the effective porosity determined by helium porosimetry, V_b is the bulk volume of the specimen determined by mercury immersion and V_g is the grain volume. Further details on the method and procedures can be found in the literature (e.g. American Petroleum Institute 1960; Luffel & Howard 1988). As described here, this method is based on the Boyle–Mariotte law. That is, a change in gas volume or gas pressure causes a commensurate change in gas pressure or volume, assuming that temperature remains constant.

Experimental results

Bulk-density and grain-density data

Bulk-density data were obtained for 33 of the samples (Table 1) by mercury porosimetry (Dorsch et al. 1996). The values are in the range of 2.63–2.77 g ml^{-1}, with an arithmetic mean of 2.71 ± 0.03 g ml^{-1}. There is a tendency for the values for samples from the Copper Creek thrust sheet to be greater than those from the Whiteoak Mountain thrust sheet (Dorsch et al. 1996). The bulk-density values for the Conasauga Group mudrock are generally greater than those commonly reported for 'shale' (Olhoeft & Johnson 1989: 2.06–2.67 g ml^{-1}; Katsube & Issler 1993), and are closer to values reported for slate (Olhoeft & Johnson 1989: 2.72–2.84 g ml^{-1}). Values as high as 2.80 g ml^{-1}, however, have been reported for tight shales, from the Venture gas field, offshore Nova Scotia (Katsube et al. 1991).

Grain-density data were obtained by helium porosimetry (Dorsch et al. 1996) for the same 33 samples (Table 1) as those for bulk density. The values are in the range of 2.70–2.83 g ml^{-1}, with a modal value of 2.79 g ml^{-1} and an arithmetic mean of 2.77 ± 0.03 g ml^{-1}. There is a tendency for the values for samples from the Copper Creek thrust sheet to be slightly greater than those from the same formation within the Whiteoak Mountain thrust sheet (Dorsch et al. 1996). In general, the grain-density values compare well with those reported for other basinal shale sequences (Katsube & Issler 1993).

In principle, it should also be possible to determine the effective porosity values using the grain- and bulk-density data in Table 1. However, these two densities were determined in two separate laboratories using different techniques (helium and mercury porosimetry), and the difference in measurement accuracies between the two techniques were not refined to the extent that such porosity determinations based on density data were possible. For example, a difference of ± 3% in measurement accuracy between the two techniques could result in a difference of 50–70% of the porosity values determined by such a method, therefore making it impossible to obtain porosity values that can be reliably compared with those determined directly with the two different porosimetry techniques.

Pore throat size distribution data

Data for the pore throat size distribution and their geometric mean were obtained for 33 of the samples (Table 1) by mercury porosimetry (Dorsch et al. 1996). The shape of the pore throat size distributions is generally unimodal (Fig. 5), similar to the results in previous studies (Katsube & Williamson 1994), with some having

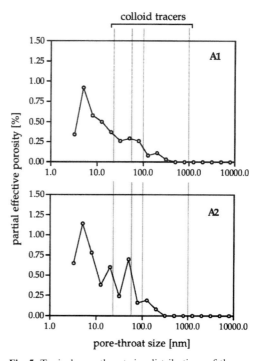

Fig. 5. Typical pore throat size distributions of the Conasauga Group mudrock from the Oak Ridge Reservation. Superimposed is the size range of colloidal tracers (diameters of 26 nm, 62 nm, 100 nm, 1000 nm) used in the tracer experiments by McKay et al. (1995).

Table 1. *Effective porosity and density (grain and bulk) data of mudrock from the Conasauga Group on the Oak Ridge Reservation*

Core	Sample	Depth (m)	ϕ_{He} (%)	ϕ_{Hg} (%)	ϕ_{I1} (%)	ϕ_{I2} (%)	d_{Hg} (nm)	δ_{He} (g ml^{-1})	δ_{Hg} (g ml^{-1})	Strat. unit
GW-133	A1	41.07	11.4	3.8	7.67*	7.08	30.5	2.73	2.64	MyLs
GW-133	A2	67.18	12.7	4.9	11.47*		24.5	2.78	2.71	MyLs
GW-133	A3	80.52	10.2	3.1	11.83*		51.3	2.74	2.73	MyLs
GW-133	A4	95.86								MyLs
GW-133	A5	114.53	7.6	3.4	11.51		33.9	2.74	2.70	MyLs
GW-133	A6	138.73	11.5	3.0	10.90*		40.7	2.72	2.67	RvSh
GW-133	A7	163.12	12.7	3.5	11.03*		19.2	2.75	2.71	RvSh
GW-133	A8	165.56	19.2	4.4	9.75		49.3	2.81	2.74	RvSh
GW-132	A9	45.95			9.16					RtLs
GW-132	A10	65.33	5.1	2.9	9.39		35.9	2.73	2.72	RtLs
GW-132	A11	90.73	9.3	3.8	9.24*		41.1	2.77	2.70	PVSh
GW-132	A12	102.97	10.7	3.0	10.35		38.7	2.76	2.72	PVSh
GW-132	A13	130.71			11.41					PVSh
GW-132	A14	130.76	6.3	4.5	9.43		26.2	2.82	2.72	PVSh
GW-132	A15	187.83	3.8	3.1	11.44		66.5	2.78	2.77	PVSh
GW-134	A16	44.45	9.9	2.7	9.46*		31.9	2.73	2.69	NcSh
GW-134	A17	58.27	12.2	3.4	11.52*		51.2	2.78	2.70	NcSh
GW-134	A18	80.29	3.2	3.8	12.04*		25.4	2.79	2.71	NcSh
GW-134	A19	99.80	2.9	4.3	13.29		42.7	2.79	2.69	NcSh
GW-134	A20	109.53	4.9	4.3	15.87		56.6	2.76	2.77	NcSh
GW-134	A21	151.59	3.9	4.0	9.16		51.9	2.79	2.70	NcSh
GW-134	A22	158.27	4.7	5.1	11.60		58.2	2.70	2.68	NcSh
GW-134	A23	171.86	14.7	4.2	11.95		38.0	2.79	2.67	NcSh
GW-134	A24	181.14	4.1	3.7	11.74		23.1	2.77	2.69	NcSh
GW-134	A25	201.19	10.4	3.2	10.57		49.8	2.80	2.67	NcSh
Wol-1	B1	12.04			13.00*					NcSh
Wol-1	B2	12.95								NcSh
Wol-1	B3	26.67	4.4	4.2		3.67	96.8	2.83	2.74	NcSh
Wol-1	B4	38.41	5.3	4.1			41.5	2.79	2.71	NcSh
Wol-1	B5	57.38	6.0	5.2	10.81*		49.1	2.82	2.72	NcSh
Wol-1	B6	81.43								NcSh
Wol-1	B7	99.90	10.9	3.2	11.80		30.9	2.77	2.71	NcSh
Wol-1	B8	243.84	15.4	3.4	7.43		54.0	2.79	2.67	MyLs
Wol-1	B9	320.09	7.8	3.5	6.84		51.1	2.79	2.74	RtLs
Wol-1	B10	352.60	3.5	3.2	5.35	5.15	50.4	2.79	2.76	PVSh
0.5MW012A	C2	51.44	3.9	3.1	12.84		18.1	2.77	2.72	MyLs
0.5MW012A	C3	148.10	3.6	4.5	6.44		27.2	2.78	2.68	RtLs
0.5MW012A	C4	83.10	11.8	4.2	4.58		23.3	2.81	2.73	RvSh
0.5MW012A	C5	118.10			9.59					RvSh
0.5MW012A	C6	38.34				5.41				MyLs
0.5MW012A	C7	135.13	3.7	4.5	7.97	5.94	33.3	2.78	2.70	RvSh

ϕ_{Hg}, mercury porosity; ϕ_{I1}, immersion porosity for irregularly shaped specimens; ϕ_{I2}, immersion porosity for cylindrically shaped specimens; ϕ_{He}, helium porosity; δ_{Hg}, bulk density (obtained by mercury porosimetry); δ_{He}, grain density (obtained by helium porosimetry); d_{Hg}, geometric mean of pore throat size distribution; Strat. Unit, stratigraphic unit; MyLs, Maryville Limestone; RvSh, Rogersville Shale; PvSh, Pumpkin Valley Shale; NcSh, Nolichucky Shale; RtLs, Rutledge Limestone.* Averaged value for several measurements.

weak secondary modes. The secondary mode is always within a larger pore throat size class. The general range of the pore-throat sizes is 3–5000 nm, with the main concentration in the range of 3–100 nm. The lower limit of 3 nm is due to the instrumentation limit. Pores and their porosities for pore-throat sizes smaller than 3 nm cannot, therefore, be characterized by this method. The geometric mean values of the pore throat size distributions are in the range of 18.1–96.8 nm (Table 1).

Effective porosity data

Immersion porosity (ϕ_I) was determined for 120 specimens, representing 37 samples and yielding 56 reliable measurements (Dorsch et al. 1996). Whereas the ϕ_I values are in the range of 3.67–

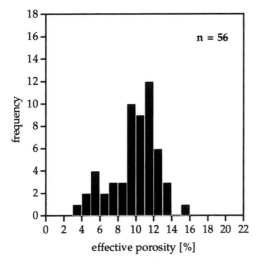

Fig. 6. Frequency distribution of immersion porosity (ϕ_I) values.

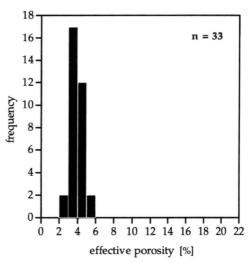

Fig. 7. Frequency distribution of mercury porosity (ϕ_{Hg}) values.

15.87% (Table 1, Fig. 6) with a mean of 9.90 (± 2.62)%, the main body is concentrated in the 9–13% range with a mode of 11–12%. A smaller secondary concentration is seen in the 4–7% range with a mode of 5–6%. The ϕ_I values for the Whiteoak Mountain thrust sheet are slightly higher (10.61 (± 1.95)%), with a smaller scatter, than those for the Copper Creek thrust sheet (8.3 (± 3.33)%).

Mercury porosity (ϕ_{Hg}) was determined for 33 samples (Table 1; Dorsch et al. 1996). The ϕ_{Hg} is represented by ϕ_{Hg1} values, which is the porosity for the pore-size range of 3 nm–10 μm, as previously indicated. The measurement accuracy is estimated to be ± 10–20%. The ϕ_{Hg} values are in the range of 3.0–5.2% (Table 1), and display a distinct unimodal distribution with a mode of 3–4% (Fig. 7). The mean value is 3.8 (± 0.7)%, with a very small scatter. These porosities are considerably smaller than those determined by the other methods.

Helium porosity (ϕ_{he}) was determined for the same 33 samples (Table 1) as those for ϕ_{Hg} (Dorsch et al. 1996). The accuracy of these measurements is estimated to be ± 10–20%. The values are in the range of 2.9–19.2% with a mean of 8.1 (± 4.3)%, but display a significant bimodal distribution (Fig. 8), with one mode at 3–4% and the other at 10–11%. The mean values for the Whiteoak Mountain and Copper Creek thrust sheets are 8.7 (± 4.4)% and 6.9 (± 4.1)% (Dorsch et al. 1996), respectively. This data set displays a considerable scatter.

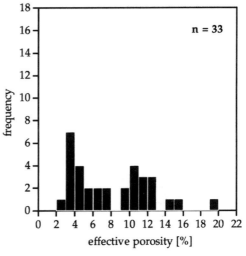

Fig. 8. Frequency distribution of helium porosity (ϕ_{he}) values.

Discussion

Comparison of effective porosity data

There is a tendency for immersion porosity (ϕ_I) to show the highest and mercury porosity (ϕ_{Hg}) to show the lowest values for an identical sample (Table 1). The helium porosity (ϕ_{he}) values show distinct divisions between high or low, depending on the sample. Four characteristic groups can be identified on the basis of a

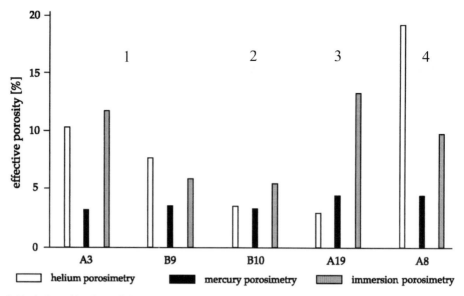

Fig. 9. Typical combinations of the three porosities (ϕ_I, ϕ_{Hg}, ϕ_{he}) representing Patterns 1–4. (A pattern refers to a typical combination of effective porosity data obtained for the identical sample by mercury, helium and immersion porosimetry.)

combination of these trends, as displayed in Figs 9 and 10. These groups will be represented by Patterns 1–4 in this study (a pattern refers to a typical combination of effective porosity data obtained for the identical sample by mercury, helium and immersion porosimetry).

Pattern 1. This group is characterized by high ϕ_I and ϕ_{he} values combined with low ϕ_{Hg} values (Figs 9 and 10), and is a combination most commonly observed in this study. A typical example is Sample A3 (Fig. 9). This combination is interpreted to represent mudrock that has considerable pore space accessible only through pore-throat sizes smaller than 3 nm, which water and helium can penetrate, but mercury cannot (Katsube 1992; Issler & Katsube 1994). These small pore throats (<3 nm) probably have been produced either by a diagenetic overprint with cement coating the surfaces, or by the well-compacted mudrock texture (Katsube & Williamson 1994). The fact that these mudrock samples yield high effective porosity values for ϕ_I and ϕ_{he}, regardless of their being well compacted, may imply that the pore space sheltered behind the small pore throats (<3 nm) is that of secondary dissolution pores.

Pattern 2. This group is characterized by congruently low values for all three porosities (ϕ_I, ϕ_{he} and ϕ_{Hg}), with Sample B10 (Fig. 9) providing

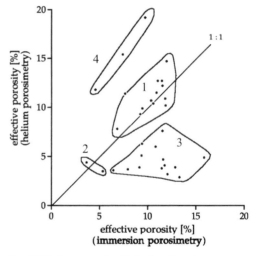

Fig. 10. Helium porosity (ϕ_{he}) plotted against immersion porosity (ϕ_I), displaying the four groups representing Patterns 1–4 (ϕ_{Hg} is uniformly low).

a typical example. This combination is interpreted to represent mudrock that lacks pore space accessed by pore throats smaller than 3 nm. Alternatively, it might indicate that many of the pore throats are blocked by diagenetic cement, which raises the possibility of significant isolated pore space in the samples.

Pattern 3. This group is characterized by values of high ϕ_I, low ϕ_{Hg} and an equal or even lower ϕ_{he}, with Sample A19 representing a typical example (Fig. 9). This combination is thought to represent possible measurement problems in obtaining the ϕ_{he} values, because $\phi_{Hg} > \phi_{he}$ trends are, otherwise, unexplainable. This is supported by the fact that Pattern 1 and 3 combinations of porosities alternate along some coreholes (e.g. in the Nolichucky Shale of the Whiteoak Mountain thrust sheet), whereas ϕ_I values remain more or less constant (Dorsch et al. 1996).

Pattern 4. This group is characterized by values of high ϕ_I and low ϕ_{Hg}, but with ϕ_{he} values considerably higher than those of ϕ_I, with Sample A8 representing a typical example (Fig. 9). This pattern is interpreted to be essentially similar to Pattern 1, except for the considerably larger ϕ_{he} values. The $\phi_I < \phi_{he}$ relationship may be explained by inhomogeneities within the mudrock sample, or by small cracks that may have developed during the ϕ_{he} measurement procedure (Dorsch et al. 1996).

Effective porosity and depth

A typical characteristic of basinal mudrock is the exponential decrease in porosity with depth, mainly through compaction (e.g. Rieke & Chilingarian 1974; Katsube & Williamson 1995). At burial depths in excess of 2 km, the decrease in porosity is small. Recent studies (e.g. Katsube & Williamson 1994, 1995) suggest that maximum compaction of mudrock is approached at the critical depth of burial, at about 2.5–3.0 km. Whereas porosity will rapidly decrease with burial depth until the critical depth of burial, it will show little decrease or may even show a slight increase with depth at burial depths greater than the critical depth of burial (Katsube & Williamson 1994, 1995). The burial depth of the Conasauga Group mudrock on the Oak Ridge Reservation is estimated to be at least 4 km (Foreman 1991; Lemiszki, pers. comm.), implying that they should have experienced maximum compaction. The shapes of the pore throat size distribution curves, furthermore, suggest that the mudrock specimens of the Conasauga Group are indeed well compacted and have experienced burial depths in excess of 2.5 km. This suggestion is based on a comparison with pore throat size distribution patterns from other basinal mudrock sequences (Katsube & Williamson 1994, 1995; Katsube et al. 1995). In addition, the distribution curves imply that the main fluid-transport pores are probably accessed and connected by pore throats in the size range of 4–60 nm. This interpretation is based on comparison of porosity studies of basinal mudrock samples with permeability measurements and formation-factor determinations on the same samples (Katsube et al. 1991, 1992b).

Effective porosity as a function of depth is shown in Fig. 11 for the immersion (ϕ_I), mercury (ϕ_{Hg}) and helium (ϕ_{he}) porosities from the three coreholes (40–200 m) in the Whiteoak Mountain thrust sheet. The depths represent present-day depth below ground surface, following structural and erosional uplift since Permian time. These porosity values are either constant or show a slight increase with increasing depth, without showing any indication of decrease with depth. These trends conform with expectations for mudrock buried to a depth of 4 km (greater than the critical depth of burial) or greater. The ϕ_{Hg} values are generally constant with depth and are uniformly lower than the ϕ_I and ϕ_{he} values, again highlighting the fact that they exclude pore space shielded by pore throats smaller than 3 nm. The ϕ_{he} values are either in agreement with or are lower than the ϕ_I values, which in some cases show a tendency to increase with increasing depth (e.g. corehole GW-132).

The same types of effective porosity (ϕ_I, ϕ_{Hg} and ϕ_{he}) v. depth curves are shown in Figs 12 and 13, for coreholes Wol-1 (0–350 m) and 0.5MW0I2A (40–150 m) in the Copper Creek thrust sheet. In contrast to the data from the Whiteoak Mountain thrust sheet (Fig. 11), these two plots display a tendency for a ϕ_I decrease with depth. This raises questions about the reason for this trend, which is for mudrock that has experienced burial depths in excess of the critical depth of burial. A possible explanation is that meteoric water has caused replacement of some of the original cement in the mudrock, and that at least part of that replaced cement has dissolved during the immersion porosity measurement procedure in the laboratory. If the degree of replacement increases toward the present-day surface, the ϕ_I is likely to also show an increase towards the surface, as displayed in Figs 12 and 13. Analysis of mudrock texture, using, for example, scanning electron microscopy, light microscopy, and X-ray diffraction, is necessary to test this hypothesis.

Implications for contaminant transport

The effective porosities obtained by this study (3.8–9.9%) for the Conasauga Group mudrock are considerably higher than previously measured (0.1–3.4%, Diment & Robertson 1963; de Laguna et al. 1968; Goldstrand pers.

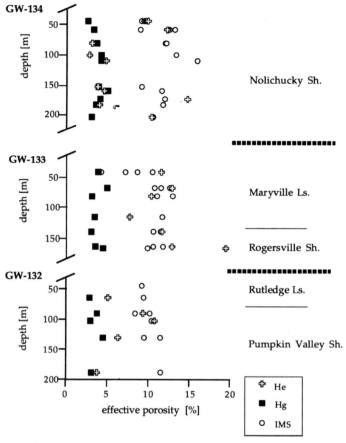

Fig. 11. Effective porosities: ϕ_I, ϕ_{Hg} and ϕ_{he}, as a function of depth for the coreholes GW-132, -133, and -134 in the Whiteoak Mountain thrust sheet. IMS, Hg and He represent ϕ_I, ϕ_{Hg} and ϕ_{he}, respectively.

comm.) or assumed (Toran et al. 1995). This discrepancy in the porosity values is probably due to different procedures used in the earlier studies (e.g. water-saturation time, vacuum saturation or not, drying procedures), which are not always described in detail. The new set of effective porosity data implies that the pore space available to influence the effect of matrix diffusion on contaminant transport within the fractured aquitard of the Conasauga Group is larger than previously known. The effect of matrix diffusion on contaminant transport includes retardation of the spread of contaminants and development of secondary contaminant sources. Therefore, the new data on porosity characteristics have important implications for the design of remediation measures. Although the effective porosity values may be somewhat smaller in principle for *in situ* conditions, because of increased confining pressure, that effect is expected to be minimal because the sample retrieval depth was <400 m (Katsube & Williamson 1994; Katsube et al. 1996).

Although, in general, the effective porosities (ϕ_I) determined by immersion porosimetry are more likely to represent the true porosity values, the possible loss of replaced meteoric cement presents a concern. Therefore, ϕ_I values for samples from the Copper Creek thrust sheet (Figs 12 and 13) should be considered as maximum values. For similar reasons, the mean ϕ_I values of 9.90% for Conasauga Group mudrock on the Oak Ridge Reservation may actually have to be considered to be slightly smaller. This is especially true for ϕ_I data from sampling intervals close to the present-day ground surface, and for similar data that lack ϕ_{he} data for independent verification.

Although the concept of replaced cement is speculative in the absence of textural data, it could have important implications if it were true. For example, will the chemical characteristics of

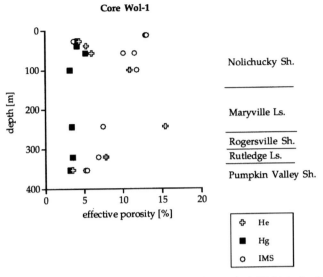

Fig. 12. Effective porosities: ϕ_I, ϕ_{Hg} and ϕ_{he}, as a function of depth for corehole Wol-1 in the Copper Creek thrust sheet. IMS, Hg and He represent ϕ_I, ϕ_{Hg} and ϕ_{he}, respectively.

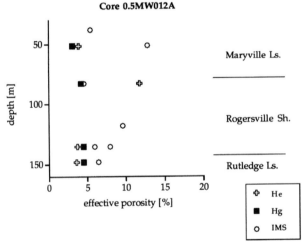

Fig. 13. Effective porosities: ϕ_I, ϕ_{Hg} and ϕ_{he}, as a function of depth for corehole 0.5MW012A in the Copper Creek thrust sheet. IMS, Hg and He represent ϕ_I, ϕ_{Hg} and ϕ_{he}, respectively.

contaminated water from the waste-disposal facilities affect the cementation of the mudrock? If they do, will they have a dissolving or enhancing effect? In addition, how would these effects influence the retardation or storage capacity of the mudrock? These are important questions that need to be addressed.

Pore size distribution data are important for evaluating the mudrock's capacity to retard the spread of contaminants, because the size of the main pore throats could determine whether the contaminant species will have access to the interconnected pore network by matrix diffusion. Whereas contaminant species smaller than the dominant pore throats would be retarded by matrix diffusion, larger species would not. For example, hydrated radii of common ions are smaller than 0.5 nm (Nightingale 1959), suggesting that they would be retarded by the Conasauga Group mudrock. Ion pairs, chelated organic compounds and colloids, however, will be larger (McCarthy & Jardine, pers. comm.). McKay et al. (1995) used colloidal tracers with molecular diameters of 26–1000 nm for microbial tracer experiments, which are unlikely to be retarded by these rocks (Fig. 5).

Pattern 2 mudrock could have a significant effect on the evaluation of contaminant retardation and storage capacity, depending on the textural source it represents. It could be dependent on three separate conditions: (a) no pore space sheltered behind the <3 nm pore throats because of a highly compacted mudrock texture, or because of an advanced stage of pore filling cementation; (b) considerable pore space exists but is isolated from the main interconnected pore network because of the <3 nm pore throats being completely blocked by diagenetic cement (advanced stage of Pattern 1); (c) same as (b) but pore throats are completely blocked by cement resulting from meteoric activities instead of earlier diagenesis. Case (c) would be an early stage of Pattern 1 if it represents meteoric cementation that had not yet been dissolved, as a result of immersion porosimetry, to the extent that the isolated pores make contact with the main interconnected pore network. Case (c) implies that there could be mudrock with retardation and storage capacities that increase with the passage of contaminated water.

Comparison of porosity-measuring techniques

The immersion porosimetry method was considered most favourable to use for the Conasauga Group mudrock, although concern existed over loss of rock material during the experimental procedure. The method was generally considered to provide accurate results, and was simple in terms of facility installation and operation of the experimental procedure. The limitations, however, were that irregular-shaped specimens cannot be used unless the bulk-density data could be obtained by other means.

Mercury porosimetry had the benefit of being able to provide important information on pore throat size distribution, which was unobtainable by the other two methods, and that irregular-shaped specimens could be used. The limitations were, however, the tendency for the effective porosity values to be underestimated for most samples, because of current instrumental limitations, and the very high cost of obtaining the results.

Helium porosimetry had the benefit of providing accurate results, in principle, for comparison with the immersion-porosimetry results, of being able to use irregular-shaped specimens, and of being relatively low in cost. The limitations were, however, the deviations in some results that were attributed to possible analytical problems. In general, all three porosimetry methods have provided a useful and powerful set of complementary techniques for porosity characterization of fine-grained siliclastic rocks of the Conasauga Group mudrock on the Oak Ridge Reservation.

Conclusions

Porosity characteristics of the Conasauga Group mudrock were determined by effective porosities measured by three techniques: immersion (ϕ_I), mercury (ϕ_{Hg}) and helium porosimetry (ϕ_{he}). The resulting porosities were in the ranges of $9.90 \pm 2.61\%$, $3.8 \pm 0.7\%$ and $8.1 \pm 4.3\%$ for ϕ_I, ϕ_{Hg} and ϕ_{he}, respectively. Bulk and grain densities, represented by values of 2.71 ± 0.03 g ml^{-1} and 2.77 ± 0.03 g ml^{-1}, respectively, were also obtained for this fractured mudrock aquitard as part of the porosity-determination procedure. This characterization was based on measurements of 37 samples for ϕ_I, and 33 samples for both ϕ_{Hg} and ϕ_{he}.

The ϕ_{Hg} values are generally the smallest of the three porosities for an identical sample. The interrelationship between the different porosities can be divided into four groups represented by Patterns 1–4: Pattern 1, ϕ_I, $\phi_{he} \gg \phi_{Hg}$; Pattern 2, low ϕ_I, ϕ_{Hg} and ϕ_{he} values; Pattern 3, $\phi_I > \phi_{Hg} \geqslant \phi_{he}$; Pattern 4: $\phi_{he} > \phi_I > \phi_{Hg}$. Patterns 1 and 2 are interpreted to result from the mudrock having particular textural characteristics, such as considerable pore space sheltered behind pore throat sizes smaller than 3 nm for Pattern 1, and either the absence of such pore space or the blockage of the <3 nm pore-throats by diagenetic cement for Pattern 2. Pattern 3 is considered to be a result of measurement problems. Pattern 4 is considered to be a result of either measurement problems or inhomogeneities within the mudrock sample.

The three porosities, generally, show little change with depths for the five coreholes (100–400 m depth range) in the two thrust sheets (Whiteoak Mountain and Copper Creek). This is a trend that would be expected, as burial depth of the Conasauga Group mudrock is at least 4 km, a depth greater than the critical depth of burial (i.e. 2.5–3.0 km). Little change of porosities is expected at depths greater than the critical depth of burial. However, a decrease in immersion porosity (ϕ_I) with depth is seen for the two coreholes in the Copper Creek thrust sheet. This trend might be explained by meteoric cement replacing the original diagenetic cement, which, in turn, was removed during the immersion porosimetry measurement procedures. This explanation, if correct, could have significant implications for the retardation and storage capacity of the mudrock, but this possible explanation needs to be investigated further.

Information on the porosity characteristics obtained by this study has important implications on contaminant transport in these rocks. The effective porosities obtained by this study (3.8–9.9%) are considerably higher than previously measured (0.1–3.4%), implying the necessity for a redesign of remediation measures. The change in available pore space will influence retardation of contaminant transport and development of secondary contaminant sources. In addition, although the concept of replaced cement is still speculative because of absence of textural data, it raises questions regarding the retardation and storage capacity of the Conasauga Group mudrock. For example, will the water from the waste-disposal facilities enhance or reduce the capacity? The small pore throat sizes (<3 nm) of these mudrock samples indicate that although contaminant species smaller than the dominant pore throats would be retarded by matrix diffusion, larger species would not.

As the three methods for determining porosity have their advantages and limitations, a combination of the three provides the most comprehensive and reliable set of data required for evaluating the effects of matrix diffusion on contaminant transport within these rocks. As an individual method, immersion porosimetry was most advantageous, mainly because of the simplicity of facility installation.

We thank J. Macquaker (University of Manchester), S. Horseman (British Geological Survey) and F. J. Pearson (Paul Scherrer Institut, Villingen) for the critical reading of the manuscript, which resulted in significant improvements. Any shortcomings, however, remain the responsibility of the authors. R. B. Dreier, P. J. Lemiszki and W. E. Sanford (Oak Ridge National Laboratory) helped with their advice while this research was accomplished. B. E. Dugan (University of Minnesota) and L. M. Tourkow (Purdue University) are thanked for their help during laboratory experiments. The support from the Oak Ridge Reservation Hydrology and Geology Group (ORRHAGS), from the Y-12 HSE&A Division (administered through W. K. Jago, Oak Ridge National Laboratory), from the X-10 Environmental Restoration Groundwater OU (administered through R. H. Ketelle, Oak Ridge National Laboratory), and from the EMSP96-Award/US-DOE (administered through L. D. McKay, University of Tennessee) is gratefully acknowledged. The research was supported in part by an appointment to the Oak Ridge National Laboratory Postdoctoral Research Associates Program (awarded to J. Dorsch) administered jointly by the Oak Ridge National Laboratory and the Oak Ridge Institute for Science and Education.

References

AGTERBERG, F. P., KATSUBE, T. J. & LEW, S. N. 1984. Statistical analysis of granite pore size distribution data, Lac du Bonnet batholith, eastern Manitoba. *Current Research, Part A. Geological Survey of Canada*, Paper, **84-1A**, 29–38.

AMERICAN PETROLEUM INSTITUTE 1960. *American Petroleum Institute Recommended Practices for Core-analysis Procedure. API Recommended Practice 40 (RP40)*. American Petroleum Institute, Washington, DC.

DELAGUNA, W. T., TAMURA, T., WARREN, H. O., STRUXNESS, E. G., MCCLAIN, W. C. & SEXTON, R. C. 1968. *Engineering Development of Hydraulic Fracturing as a Method for Permanent Disposal of Radioactive Wastes*. Oak Ridge National Laboratory **4259**.

DIMENT, W. H. & ROBERTSON, E. C. 1963. Temperature, thermal conductivity, and heat flow in a drilled hole near Oak Ridge, Tennessee. *Journal of Geophysical Research*, **68**, 5035–5047.

DORSCH, J. 1995. *Determination of Effective Porosity of Mudrocks — a Feasibility Study*. Oak Ridge National Laboratory–Groundwater Program Office **019**.

——, KATSUBE, T. J., SANFORD, W. E., DUGAN, B. E. & TOURKOW, L. M. 1996. *Effective Porosity and Pore-Throat Sizes of Conasauga Group Mudrock: Application, Test and Evaluation of Petrophysical Techniques*. Oak Ridge National Laboratory–Groundwater Program Office **021**.

FOREMAN, J. L. 1991. *Petrologic and geochemical evidence for water–rock interaction in the mixed carbonate–siliciclastic Nolichucky Shale (Upper Cambrian) in East Tennessee*. PhD thesis, University of Tennessee, Knoxville.

GERMAIN, D. & FRIND, E. O. 1989. Modelling of contaminant migration in fracture networks: effects of matrix diffusion. *Proceedings, International Symposium on Contaminant Transport in Groundwater, Stuttgart, Germany*, April 1989.

HASSON, K. O. & HAASE, S. C. 1988. Lithofacies and paleogeography of the Conasauga Group (Middle and Late Cambrian) in the Valley and Ridge province of east Tennessee. *Geological Society of America Bulletin*, **100**, 234–246.

HATCHER, R. D., JR, LEMISZKI, P. J., DREIER, R. B., *et al.* 1992. *Status Report on the Geology of the Oak Ridge Reservation*. Oak Ridge National Laboratory **TM-12074**.

ISSLER, D. R. & KATSUBE, T. J. 1994. Effective porosity of shale samples from the Beaufort-MacKenzie Basin, northern Canada. *Current Research, 1994-B*. Geological Survey of Canada, 19–26.

KATSUBE, T. J. 1992. Statistical analysis of pore-size distribution data of tight shales from the Scotian Shelf. *Current Research, Part E*. Geological Survey of Canada, Paper, **92-1E**, 365–372.

—— & BEST, M. E. 1992. Pore structure of shales from the Beaufort–Mackenzie Basin, Northwest Territories. *Current Research, Part D.* Geological Survey of Canada, Paper, **92-1E**, 157–162.

—— & ISSLER, D. R. 1993. Pore-size distributions of shales from the Beaufort–Mackenzie Basin, northern Canada. *Current Research, Part E.* Geological Survey of Canada, Paper, **93-1E**, 123–132.

—— & MARESCHAL, M. 1993. Petrophysical model of deep electrical conductors; graphite lining as a source and its disconnection due to uplift. *Journal of Geophysical Research*, **98**(B5), 8019–8030.

—— & SCROMEDA, N. 1991. Effective porosity measuring procedure for low porosity rocks. *Current Research, Part E.* Geological Survey of Canada, Paper, **91-1E**, 291–297.

—— & WALSH, J. B. 1987. Effective aperture for fluid flow in microcracks. *International Journal of Rock Mechanics and Mining Sciences and Geomechanics Abstracts*, **24**, 175–183.

—— & WILLIAMSON, M. A. 1994. Effect of shale diagenesis on shale nano-pore structure and implications for sealing capacity. *Clay Minerals*, **29**, 451–461.

—— & —— 1995. Critical depth of burial of subsiding shales and its effect on abnormal pressure development. *Proceedings of the Oil and Gas Forum '95 ('Energy from Sediments')*. Geological Survey of Canada, Open File, **3058**, 283–286.

——, BLOCH, J. & ISSLER, D. R. 1995. Shale pore structure evolution under variable sedimentation rates in the Beaufort–Mackenzie Basin. *Proceedings of the Oil and Gas Forum '95 ('Energy from Sediments')*. Geological Survey of Canada, Open File, **3058**, 211–215.

——, ISSLER, D. R. & COYNER, K. 1996. Petrophysical characteristics of shale from the Beuafort–Mackenzie Basin, northern Canada: permeability, formation factor, and porosity versus pressure. *Current Research 1996-B.* Geological Survey of Canada, Paper, **96-B**, 45–50.

——, MUDFORD, B. S. & BEST, M. E. 1991. Petrophysical characteristics of shales from the Scotian Shelf. *Geophysics*, **56**, 1681–1689.

——, SCROMEDA, N. & WILLIAMSON, M. 1992a. Effective porosity from tight shales from the Venture gas field, offshore Nova Scotia. *Current Research, Part D.* Geological Survey of Canada, Paper, **92-1D**, 111–119.

——, WILLIAMSON, M. & BEST, M. E. 1992b. Shale pore structure evolution and its effect on permeability. In: *Symposium Volume III of the 33rd-Annual Symposium of the Society of Professional Well Log Analysts (SPWLA)*. Society of Core Analysts Preprints, **SCA-6214**, 1–24.

KOPASKA-MERKEL, D. C. 1988. New applications in the study of porous media: determination of pore-system characteristics on small fragments (part I). *Northeastern Environmental Science*, **7**, 127–142.

—— 1991. *Analytical Procedure and Experimental Design for Geological Analysis of Reservoir Heterogeneity using Mercury Porosimetry*. Alabama Geological Survey Circular, **153**.

LOMAN, J. M., KATSUBE, T. J., CORREIA, J. M. & WILLIAMSON, M. A. 1993. Effect of compaction on porosity and formation factor for tight shales from the Scotian Shelf, offshore Nova Scotia. *Current Research, Part E.* Geological Survey of Canada, Paper, **93-1E**, 331–335.

LUFFEL, D. L. & HOWARD, W. E. 1988. Reliability of laboratory measurement of porosity in tight gas sands. *Society of Petroleum Engineers Formation Evaluation*, December, 705–710.

MARKELLO, J. R. & READ, J. F. 1981. Carbonate ramp-to-deeper shale shelf transitions of an Upper Cambrian intrashelf basin, Nolichucky Formation, southwest Virginia Appalachians. *Sedimentology*, **28**, 573–597.

—— & —— 1982. Upper Cambrian intrashelf basin, Nolichucky Formation, southwest Virginia Appalachians. *Bulletin, American Association of Petroleum Geologists*, **66**, 860–878.

MCKAY, L. D., GILLHAM, R. W & CHERRY, J. A. 1993. Field experiments in a fractured clay till 2. Solute and colloid transport. *Water Resources Research*, **29**, 3879–3890.

——, SANFORD, W. E., STRONG-GUNDERSON, J. & DE ENRIQUEZ, V. 1995. Microbial tracer experiments in a fractured weathered shale near Oak Ridge, Tennessee. *Proceedings, International Association of Hydrogeologists, Solutions '95 Conference, Edmonton, Alta.*

MONTANEZ, I. P. 1994. Late diagenetic dolomitization of Lower Ordovician, Upper Knox Carbonates: a record of the hydrodynamic evolution of the Southern Appalachian basin. *Bulletin, American Association of Petroleum Geologists*, **78**, 1210–1239.

MUSSMAN, W. J., MONTANEZ, I. P. & READ, J. F. 1988. Ordovician Knox paleokarst unconformity, Appalachians. In: JAMES, N. P. & CHOQUETTE, P. W. (eds) *Paleokarst*. Springer, Heidelberg, 211–228.

NERETNIEKS, I. 1980. Diffusion in the rock matrix: an important factor in radionuclide retardation? *Journal of Geophysical Research*, **85**, 4379–4397.

NIGHTINGALE, E. R. 1959. Phenomenological theory of ion solvation. *Journal of Physical Chemistry*, **63**, 1381–1387.

NUR, A. & SIMMONS, G. 1969. The effect of saturation on velocity in low porosity rocks. *Earth and Planetary Science Letters*, **7**, 183–193.

OLHOEFT, G. R. & JOHNSON, G. R. 1989. Densities of rocks and minerals. In: CARMICHAEL, R. S. (ed.) *Practical Handbook of Physical Properties of Rocks and Minerals.* CRC Press, Boca Raton, FL, 141–176.

READ, J. F. (1989). Controls on evolution of Cambrian–Ordovician passive margin, U.S. Appalachians. In: CREVELLO, P. D., WILSON, J. L., SARG, J. F. & READ, J. F. (eds) *Controls on Carbonate Platform and Basin Development*. Society of Economic Paleontologists and Mineralogists, Special Publication, **44**, 147–165.

RIEKE, H. H., III & CHILINGARIAN, G. V. (eds) 1974. *Compaction of Argillaceous Sediments*. Developments in Sedimentology 16. Elsevier, Amsterdam.

RODGERS, J. 1953. *Geologic map of East Tennessee with Explanatory Text, Scale 1:125 000.* Bulletin, Tennessee Division of Geology, **55**.

ROOTARE, H. M. 1970. A review of mercury porosimetry. *Perspectives of Powder Metallurgy*, **5**, 225–252.

SANFORD, W. E., JARDINE, P. M. & SOLOMON, D. K. 1994. Examining matrix diffusion in fractured shales with noble gases. *Geological Society of America, Abstracts with Programs*, **26**, A362.

SCROMEDA, N. & KATSUBE, T. J. 1993. Effect of vacuum-drying and temperature on effective porosity determination for tight rocks. *Current Research, Part E.* Geological Survey of Canada, Paper, **93-1E**, 313–319.

—— & —— 1994. Effect of temperature on drying procedures used in porosity measurements of tight rocks. *Current Research, Part E.* Geological Survey of Canada, Paper, **94-1E**, 283–289.

SHEVENELL, L. M., MOORE, G. K. & DREIER, R. B. 1994. Contaminant spread and flushing in fractured rocks near Oak Ridge, Tennessee. *Ground Water Monitoring and Remediation*, **14**, 120–129.

SOEDER, D. J. 1988. Porosity and permeability of eastern Devonian gas shales. *Society of Petroleum Engineers Formation Evaluation*, **3**, 116–138.

SOLOMON, D. K., MOORE, G. K., TORAN, L. E., DREIER, R. B. & MCMASTER, W. M. 1992. *A Hydrologic Framework for the Oak Ridge Reservation.* Oak Ridge National Laboratory **TM-12026**.

SRINIVASAN, K. & WALKER, K. R. 1993. Sequence stratigraphy of an intrashelf basin carbonate ramp to rimmed platform transition: Maryville Limestone (Middle Cambrian), southern Appalachians. *Geological Society of America Bulletin*, **105**, 883–896.

TANG, D. H., FRIND, E. O. & SUDICKY, E. A. 1981. Contaminant transport in fractured porous media: analytical solution for a single fracture. *Water Resources Research*, **17**, 555–564.

TORAN, L., SJOREEN, A. & MORRIS, M. 1995. Sensitivity analysis of solute transport in fractured porous media. *Geophysical Research Letters*, **22**, 1433–1436.

WALKER, K. R., FOREMAN, J. L. & SRINIVASAN, K. 1990. The Cambrian Conasauga Group of eastern Tennessee: a preliminary general stratigraphic model with a more detailed test for the Nolichucky Formation. *Proceedings, Appalachian Basin Industrial Associates*, **17**, 184–189.

WARDLAW, N. C. 1976. Pore geometry of carbonate rocks as revealed by pore casts and capillary pressure. *Bulletin, American Association of Petroleum Geologists*, **60**, 245–257.

——, MCKELLAR, M. & YU, LI 1988. Pore and throat size distribution determined by mercury porosimetry and by direct observation. *Carbonates and Evaporites*, **3**, 1–15.

WASHBURN, E. W. 1921. Note on a method of determining the distribution of pore sizes in a porous material. *Proceedings of the National Academy of Sciences*, **5**, 115–116.

WICKLIFF, D. S., SOLOMON, D. K. & FARROW, N. D. 1991. *Preliminary Investigation of Processes that affect Source Term Identification.* Oak Ridge National Laboratory **ER 59**.

Preferential pathways in an Eocene clay: hydrogeological and hydrogeochemical evidence

K. WALRAEVENS[1,2] & J. CARDENAL[3,4]

[1] *University of Ghent, Laboratory for Applied Geology and Hydrogeology, Krijgslaan 281-S8, 9000 Ghent, Belgium*
[2] *Fund for Scientific Research, Flanders, Belgium*
[3] *Universidad de Granada, Instituto del Agua, c/ Rector Lopez Argüeta s/n, 18071 Granada, Spain*
[4] *Present address: Universidad de Jaén, Departamento de Ingeniería Cartográfica, Geodésica y Fotogrametría, Avenida de Madrid 35, 23071 Jaén, Spain*

Abstract: This study was initiated to test the extent to which the cation exchange capacity of marine clay-rich units, through which fresh water has been recharged, controls the composition of pore waters in an aquifer. The semi-confined Ledo-Paniselian (Eocene) aquifer in Flanders is recharged in elevated regions, where it is covered by Bartonian Clay. These sediments were initially laid down in marine conditions. The effects of recharging fresh $CaHCO_3$ water by downward flow through the overlying clay within the aquifer were modelled using PHREEQM. In the aquifer, progressively fresh waters are found in an upstream direction, which indicates a reduced influence of the initial marine conditions. Different stages of cation exchange produce a chromatographic sequence of groundwater types. Hydrogeological and hydrogeochemical evidence suggests the existence of preferential, faster flow pathways for recharge groundwater flow through the Bartonian Clay overlying the aquifer. These conclusions are indicated by differences between the observed and predicted features of the aquifer. It is necessary to invoke hydraulic conductivity values of 10^{-9} m s^{-1} rather than 10^{-10} m s^{-1} (measured values) for the Bartonian Clay so that the predicted head distributions in the aquifer match the observed heads. Furthermore, the measured exchangeable cations in clay collected from cores in the recharge area show less freshening than is predicted from the model to produce either the observed groundwater type within the aquifer in the recharge area or the overall distribution of groundwater types in the aquifer. Given the discrepancy between the observed and calculated state of freshening it is assumed that the analysed flow line in the core must be part of a slower flow pathway.

The objective of this study is to test (a) the extent to which the cation exchange capacity of marine clay-rich units, through which fresh water is recharged, has controlled the composition of the pore waters in a freshening aquifer and (b) whether there are preferential flow pathways through the sediments confining the aquifer. Previous studies of aquifers, e.g. in the coastal plains in the USA (Foster 1950; Back 1966; Chapelle & Knobel 1983; Appelo 1994), in polder areas of the Netherlands (Appelo & Willemsen 1987; Beekman 1991; Stuyfzand 1993) and in Tertiary sediments in Flanders, Belgium (Walraevens 1987, 1990), have shown that the displacement of interstitial marine water by infiltration of $CaHCO_3$ water commonly results in $NaHCO_3$ water upstream of the fresh-water–salt-water interface. Moreover, the sequential release of Na^+, K^+ and Mg^{2+} from the exchange sites, in accordance with the affinity sequence, can result in the development of a chromatographic pattern, leading to the successive appearance of $NaHCO_3$, $Na(K)HCO_3$, $MgHCO_3$ and finally $CaHCO_3$ waters. This pattern has also been demonstrated in the laboratory by column experiments (e.g. Appelo et al. 1990; Beekman & Appelo 1990; Beekman 1991) and by field tests over short time intervals (Valocchi et al. 1981).

In natural aquifers, the predicted chromatographic pattern may be difficult to detect, as complex flow patterns within 'real' systems can cause mixing, which alters the predicted cation concentration gradients. Nevertheless, in spite of mixing, indications of the theoretical chromatographic patterns have been described in the

Aqua aquifer in Maryland, USA (Appelo 1994), in the Dutch polders (Beekman 1991; Stuyfzand 1993) and in the Tertiary Ledo-Paniselian aquifer in Flanders (Walraevens 1987, 1990). The first attempts to model the last pattern were described by Cardenal & Walraevens (1994) and Walraevens & Cardenal (1994). In the present study, flow patterns in the sandy Ledo-Paniselian aquifer were modelled to determine how well the natural patterns can be replicated and if there are preferential flow patterns through the clay overlying the aquifer.

The Ledo-Paniselian aquifer is semi-confined in the northern part of East and West Flanders (Fig. 1) and recharge, by infiltration through the overlying Bartonian Clay, occurs in the highest regions in the south of the survey area (Walraevens 1988).

Geological setting

The Eocene, marine Ledo-Paniselian sediments comprise the Gent Formation, the Knesselare Formation and the Lede Formation. They form part of a succession of alternating clayey and sandy, subhorizontal Tertiary deposits that dip gently ($<1°$) to the NNE. In the study area the depth to the top of the aquifer increases towards the NNE and the geology of the Tertiary surface is shown in Fig. 1. The Ledo-Paniselian Sands are overlain by clays of Upper Eocene, Bartonian age, which form part of the Maldegem Formation. These strata consist mainly of stiff clay, with some intercalations of sandy clay, and semi-confine the Ledo-Paniselian aquifer. Overlying these clays are sands of Oligocene age (Zelzate Formation), clays of Rupelian age (Boom Formation), and finally Miocene and Pliocene sands of various undifferentiated lithostratigraphic units.

The Ledo-Paniselian Sands are successively underlain by the Paniselian Clay of the Gent Formation, the Ypresian Sands of the Tielt Formation, and ultimately by a thick succession of stiff clays (Ypresian Clay of the Kortrijk Formation, Lower Eocene). The last unit is the lateral equivalent of the London Clay and is considered to be the base of the groundwater reservoir (Walraevens 1987).

Modelling of groundwater flow

The groundwater flow in the Ledo-Paniselian aquifer was calculated by means of a quasi-3D finite-difference model (Walraevens 1988). The model mesh is given in Fig. 2. The grid is composed of cells with a dimension of 2 km × 1 km. The following boundary conditions have been adopted:

(1) at the southern boundary, where the Ledo-Paniselian aquifer becomes phreatic, the hydraulic heads are assumed to be fixed;
(2) the northern boundary is assumed to be impermeable, as head measurements in that region indicate a very slow east–west orientation of the groundwater flow;
(3) the eastern boundary is divided in two parts: (a) to the north of Antwerp, fixed heads are based upon local measurements; (b) south of Antwerp the boundary is assumed to be impermeable, as the topography and groundwater movement on both sides are considered to be symmetrical;
(4) the water table over the whole area has been modelled assuming fixed hydraulic heads;
(5) the base of the reservoir is considered to be impervious as it rests upon the thick Ypresian Clay.

Pumping tests have allowed reliable hydraulic conductivity data to be obtained from the sands. In contrast, however, it was found to be very difficult to obtain reliable conductivity data for the clay-dominated sediments. Laboratory tests indicate that conductivity values for these clays are either $c.\ 10^{-10}$ m s^{-1} or lower (Walraevens 1987). Using this value for the conductivity of the clay and field measurements of the observed hydraulic head distribution in the aquifer and at the water table (well known in the westerly recharge area around Ursel, as data from 56 piezometers were available), the predicted head distribution was calculated assuming that the aquifer is semi-confined. The observed hydraulic head below the Bartonian Clay, however, could not be simulated assuming a vertical hydraulic conductivity of 10^{-10} m s^{-1} for the clay, as the calculated heads were much too low. Better hydraulic conductivity predictions were made by calibrating the model using the observed hydraulic head distribution. Using this approach, an excellent agreement between observed and calculated heads was obtained assuming a value of 10^{-9} m s^{-1} for the hydraulic conductivity of the clay. The discrepancy between the hydraulic conductivity values obtained from laboratory tests on local samples and the values deduced from integrating measured regional hydraulic head data has been ascribed to the presence of preferential pathways in the clay, through which flow is preferentially taking place (Walraevens 1987).

Two important recharge areas of the aquifer have been delimited by means of this model: one in the west around Ursel, and one in the east,

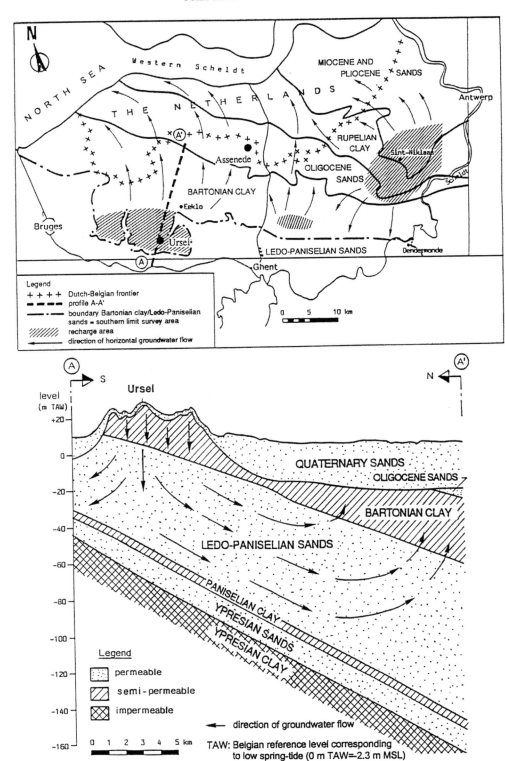

Fig. 1. Geology of the survey area and groundwater flow.

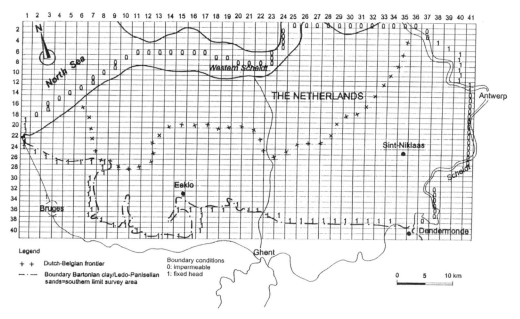

Fig. 2. Model grid and boundary conditions of hydraulic model.

around Sint-Niklaas (shaded in Fig. 1). These areas correspond to topographically elevated regions. A third, relatively minor, recharge area has also been observed between these two regions. In the aquifer, close to the recharge areas, the hydraulic heads decrease to the north.

The location of discharge areas in the Ledo-Paniselian aquifer has been constrained by historic hydraulic head observations. For instance, over large parts of the discharge areas, in the polders where the topographic elevation is low, the wells used to be artesian. Recent groundwater extraction, however, has substantially changed the flow pattern in the aquifer. In spite of recent extraction, it is the pre-extraction flow pattern that determines the distribution of the different chemical groundwater types (Walraevens 1990).

The calculated groundwater flow in natural conditions shows that the aquifer is recharged by infiltration through the Bartonian Clay. This is schematically represented in cross-section A–A' in Fig. 1. From the recharge areas, groundwater generally flows towards the north to northwest. Outside the recharge areas, a very slow upward flow occurs, along which groundwater leaves the aquifer by flowing out through the Bartonian Clay. This vertical outflow gradually reduces the velocity within the aquifer, along the flow direction (Walraevens 1988).

Hydrogeochemical evolution

The studied Tertiary sediments were all deposited in a marine environment. At the end of Tertiary time, relative sea level fell and the present-day topography developed as a result of fluvial erosion (Tavernier 1954). Groundwater movement was induced by precipitation recharge in more elevated regions. In the recharge areas calcite dissolved in the infiltrating water. Freshening of the aquifer started, expressed both by dilution of the marine pore waters and by cation exchange (with the latter predominantly occurring in the Bartonian Clay; Walraevens 1990). The early marine groundwater was pushed to the north by the infiltrating fresh water and the fluid movement in the aquifer.

The modern groundwater quality distribution in the Ledo-Paniselian aquifer is interpreted to have been generated by this process (Fig. 3; groundwater classification according to Stuyfzand (1986)). For instance, the measured variation of groundwater composition along profile A–A' (see Fig. 1), shows that salinity increases in the downstream direction, as a result of the decreasing effects of dilution downstream from the recharge area. For instance, in the recharge area, the groundwater is fresh (symbol F: <150 mg l^{-1} Cl$^-$); further north it is fresh–brackish (symbol F_b: 150–300 mg l^{-1} Cl$^-$);

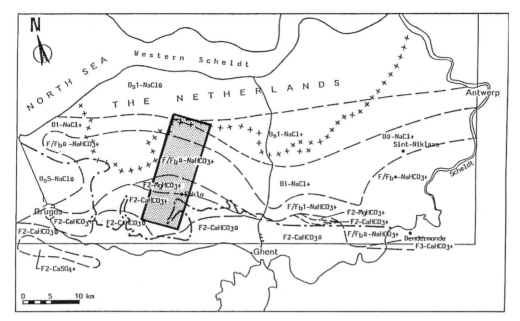

Fig. 3. Distribution of groundwater types in the Ledo-Paniselian aquifer (groundwater classification after Stuyfzand 1986).

further north still it is brackish (symbol B: 300–1000 mg l^{-1} Cl$^-$); finally, in the north of the study area it is brackish–salt (symbol B$_S$: 1000–10 000 mg l^{-1} Cl$^-$). Moreover, along the same profile, the groundwater type changes from south to north: i.e. from CaHCO$_3$ in the southwesterly recharge area around Ursel to, sequentially, MgHCO$_3$, NaHCO$_3$ and finally NaCl in the north of the study area. The two intermediate water types have been generated from the two end members by cation exchange, and the chromatographic sequence of cation exchange is expressed upgradient (progressively fresher water) by subsequent surplus of Na$^+$, followed by K$^+$ and finally Mg^{2+}, resulting in the NaHCO$_3$ and MgHCO$_3$ water types, respectively. The groundwater leaking out of the clay and entering the aquifer nowadays in the recharge area contains only the Ca^{2+} cation in appreciable quantities. Thus, the sequence of groundwater types observed in the Ledo-Paniselian aquifer is in excellent agreement with the pattern of natural groundwater flow.

Cores from a borehole in the recharge area at Ursel have provided the opportunity to determine the cation exchange capacity (CEC) and exchangeable cations of both the clay and the underlying aquifer (Walraevens & Lebbe 1988). The determination of the CEC and the extraction of the exchangeable cations were performed by means of NH$_4^+$-acetate at pH 7. In this analytical method, NH$_4^+$ replaces the adsorbed cations, and the original adsorbed cations are then analysed. As salts soluble in water (e.g. Na–Cl, Ca–SO$_4$) also dissolve in NH$_4^+$-acetate, they were determined separately, and then subtracted from the NH$_4^+$-acetate results. The CEC for the Bartonian Clay at Ursel, where the clay content is 70% and the silt content is 30%, is c. 27 mequiv per 100 g. In contrast, the CEC in the Ledo-Paniselian aquifer, where the clay content averages 7% and the silt content 3%, is 5.5 mequiv per 100 g. The exchangeable cations in the Bartonian Clay at Ursel were determined to be (Table 1): CaX$_2$ = 12 mequiv per 100 g, ranging between 5 and 18 mequiv per 100 g; MgX$_2$ = 13 mequiv per 100 g, ranging between 7 and 19 mequiv per 100 g; and <3 and 0.2 mequiv per 100 g for KX and NaX, respectively (Walraevens & Lebbe 1988).

The Bartonian Clay at Ursel shows a very clear depletion in adsorbed Na$^+$, when compared with a borehole outside the recharge area, at Assenede (Table 1). Here adsorbed Ca^{2+} is lower than at Ursel. The distribution of adsorbed cations confirms that the Bartonian Clay has been significantly leached in the recharge area. In the discharge area, the marine influence still persists, and is expressed by a high concentration of NaX.

The analysed samples of the Bartonian Clay at Ursel within the recharge area, however, do not show a depletion in adsorbed Mg^{2+} and K$^+$.

Table 1. *Amounts of exchangeable cations (in mequiv per 100 g) in the Bartonian Clay in the recharge area at Ursel and in the discharge area at Assenede (after Walraevens 1987)*

	Ursel			Assenede		
	Mean	Range	n	Mean	Range	n
CaX_2	12	5–18	15	11	7–16	18
MgX_2	13	7–19	15	10	5–17	18
NaX	0.2	0.1–0.2	15	3.9	0.5–9.0	18
KX	1.9	0.4–3.0	14	1.8	0.2–3.2	18

n, number of clay samples analysed.

This is surprising, as the Ledo-Paniselian aquifer in the recharge area contains $CaHCO_3$ water, which suggests that flushing of the overlying clay layer is almost complete (Walraevens & Cardenal 1994). This arises because the infiltration water does not contain appreciable amounts of Mg^{2+} or K^+, and the original interstitial solution contains almost no Ca^{2+}. In this setting the affinity sequence would be expected to release first Na^+ and K^+, and finally Mg^{2+} from the clay surfaces in exchange for Ca^{2+}, so that the produced groundwater will contain Ca^{2+} as the main cation only after the marine cations have largely flushed out of the sediment. In the study area, although the Na^+ has been removed, adsorbed K^+ and Mg^{2+} are still present in high concentrations. Moreover, it might be expected that as water infiltrates the aquifer by percolating through a clay with Mg^{2+} as its main adsorbed cation, the water would exchange part of its Ca^{2+} for Mg^{2+}. This process would then lead to the groundwater being enriched in Mg^{2+} and depleted in Ca^{2+}. This pattern, however, is not observed in the recharge area, where the waters are of a $CaHCO_3$ type; thus a discrepancy exists between the cations adsorbed on the overlying clay and the groundwater composition of the aquifer.

Cation exchange in the geochemical model PHREEQM

Ion exchange can be described as a reaction where an equilibrium constant is the exchange coefficient (Appelo & Postma 1993). For the general reaction

$$Na^+ + 1/iIX_i \rightleftharpoons NaX + 1/iI^{i+}$$

where X represents the cation exchanger.

Using the Gaines–Thomas convention for the activities of adsorbed cations,

$$K_{Na/I} = \frac{\beta_{Na}[I^{i+}]^{1/i}}{\beta_I^{1/i}[Na^+]}$$

where the equivalent fraction β_I for ion I^{i+} is calculated as

$$\beta_I = \frac{\text{mequiv } IX_i \text{ per 100 g sediment}}{CEC}.$$

The exchange coefficients are expressed with Na^+ as the reference cation. Selecting an arbitrary high value (10^{20}) for the association constant, K_{NaX}, which corresponds to the half-reaction

$$Na^+ + X^- \rightleftharpoons NaX$$

$$K_{NaX} = \frac{\beta_{Na}}{[Na^+][X^-]}$$

we can define the association constant K_{IX_i} for the other half-reaction,

$$I^{i+} + iX^- \rightleftharpoons IX_i$$

in terms of the association constant K_{NaX} and the exchange coefficient $K_{Na/I}$:

$$K_{IX_i} = \left(\frac{K_{NaX}}{K_{Na/I}}\right)^i$$

(Appelo & Postma 1993).

The association constants K_{IX_i} for the different adsorbed cations can be calculated on the basis of data for the equivalent fraction β_I of the adsorbed cation, and the cation's pore-water concentration:

$$K_{IX_i} = \frac{\beta_I}{[I^{i+}][X^-]^i}$$

in which $[X^-]$ is replaced by $\beta_{Na}/[Na^+]K_{NaX}$.

Hydrogeochemical modelling

The geochemical–mixing cell model PHREEQM (Appelo & Postma 1993) has been used to simulate the freshening of the Bartonian Clay (Walraevens & Cardenal 1994) and the subsequent recharge to the underlying aquifer (Cardenal & Walraevens 1994). The modelling was applied to the cross-section A–A' through the westerly recharge area at Ursel, represented in Fig. 1. A flow line was selected, with its origin at the top of the Bartonian Clay, which was directed towards the NNE within the aquifer. Model results have been compared with groundwater analyses of the aquifer derived from pumped samples taken from the rectangular area indicated in Fig. 3. Modelling suggests that the present groundwater type distribution in the Ledo-Paniselian aquifer mainly results from cation exchange processes that occurred in the overlying Bartonian Clay and, to a lesser extent, within the aquifer. In addition, the presence of some oxygen within the groundwater has been postulated, as the pyrite weathering product gypsum has been observed. Moreover, the effects of organic matter to fuel sulphate reduction have been modelled and the influence of the calcite precipitation–dissolution processes has been taken into account.

Groundwater flow conditions in the Bartonian Clay differ from those in the Ledo-Paniselian aquifer, with respect to flow velocity and length of flow path. For instance, the clay is relatively thin and the flow path through it is relatively short (tens of metres); in contrast, the aquifer is much thicker and the flow path through it is relatively long (tens of kilometres). Therefore, the modelling process was divided in two parts. First, the freshening of the clay was simulated in a vertical column. Second, the effects of varying water types leaking out of the Bartonian Clay into the Ledo-Paniselian aquifer were simulated by flushing the varying water types produced from the clay through a horizontal column to simulate the effects of water movement in the main aquifer.

An overview of the model boundaries is given in Table 2. For both clay and aquifer, the initial pore water quality was sea water, equilibrated with calcite and 0.75 mmol kg^{-1} H$_2$O of NH$_4^+$ production. This solution was also equilibrated with the exchange complex in the Bartonian Clay and in the Ledo-Paniselian aquifer, respectively. The composition of the recharge water was assumed to be pure water equilibrated with calcite and a P_{CO_2} of $10^{-2.0}$ atm. The flow velocities were deduced from the flow model (Walraevens 1988). The cation exchange variables for the Bartonian Clay were computed from measured pore-water compositions and data for adsorbed cations taken from Walraevens (1987). For the Ledo-Paniselian aquifer, the standard coefficients in the database of PHREEQM were used. The similarity between computed and standard coefficients gives confidence in the reliability of the exchange model.

The evolution of the distribution of exchangeable cations adsorbed to the Bartonian Clay, and of the composition of the pore water leaking out of it during the flushing, are shown in Fig. 4. After the clay has been flushed 100 times, the distribution on the clay's adsorption complex is similar to that found in the sediments from the Ursel core. At this point in the model, the pore-water solution leaving the clay contains mainly Mg^{2+} and Ca^{2+}, with no significant Na$^+$. This computed composition, however, is very different from the one found at present in the Ledo-Paniselian aquifer in the recharge area of Ursel, where Mg^{2+} is virtually absent. To match the computed water type with that observed in the aquifer, the clay must be flushed nearly 400 times. It is thus likely that the aquifer must have been recharged via additional pathways to the one we investigated in the Ursel core. This strongly suggests that there are preferential pathways through the Bartonian Clay, where the flow rates are faster than the flow line we analysed. Furthermore, the region we analysed must be part of a slower flow system, where freshening is still in progress (Walraevens & Cardenal 1994).

When solutions of varying concentrations, produced after flushing the clay 100 times, were used to flush the Ledo-Paniselian aquifer, the modelled groundwater quality distributions were found to be very different from the observed groundwater distributions (Fig. 5). The clay had to be flushed nearly 400 times to obtain a close match between the observed and computed groundwater distributions (Fig. 6). These observations also suggest that there are preferential fast fluid flow lines through the Bartonian Clay.

Cation exchange mainly occurred in the Bartonian Clay during recharge. Some, however, occurred in the aquifer. The first and most significant (in absolute concentrations) cation exchange reaction is the desorption of Na$^+$. This process led to an NaHCO$_3$ water type. This water type, which obtained its main characteristics during infiltration through the Bartonian Clay, has been displaced by groundwater flow. It is currently occurring between 14 and 6 km from the recharge area, according to the model results. Beyond 14 km the sea-water–fresh-water mixture

Table 2. *Model boundaries for freshening of Bartonian Clay and Ledo-Paniselian aquifer*

Hydraulic and hydrochemical conditions for modelling the freshening of the Bartonian Clay and Ledo-Paniselian aquifer

Infiltration water
$Ca^{2+} = 2.1$ mmol l^{-1}
$HCO_3^- = 4.17$ mmol l^{-1}
pH = 7.3
$\log P_{CO_2} = -2.00$

Initial pore water (sea water equilibrated with calcite and 0.75 mmol l^{-1} NH_4^+ production)
$Ca^{2+} = 9.8$ mmol l^{-1} $K^+ = 10.6$ mmol l^{-1} $Cl^- = 566$ mmol l^{-1}
$Mg^{2+} = 55.1$ mmol l^{-1} $SO_4^{2-} = 29.3$ mmol l^{-1} $NH_4^+ = 0.75$ mmol l^{-1}
$Na^+ = 486$ mmol l^{-1} $HCO_3^- = 1.6$ mmol l^{-1} pH = 7.86

(1) Bartonian Clay
Column length: 20 m
Number of cells: 4 (5 m each)
Flow velocity: 0.12 m year^{-1}
Time step: 41.7 years
Dispersivity: 1.5 m
Diffusion coefficient: 10^{-9} m^2 s^{-1}
Cation exchange capacity: 27 mequiv per 100 g
Gypsum dissolution: 0.45 mmol l^{-1} per cell
Calcite equilibrium

Exchange parameters (computed from data of Walraevens 1987)
$\log K_{NaX} = 20.0$ (reference value)
$\log K_{KX} = 20.6$
$\log K_{CaX_2} = 41.2$
$\log K_{MgX_2} = 40.9$
$\log K_{NH_4X} = 20.6$ (from Appelo & Postma 1993)
$\log K_{HX} = 22.5$ (Appelo, pers. comm.)

(2) Ledo-Paniselian aquifer
Column length: 20 km
Number of cells: 20 (1 km each)
Flow velocity: 3.35 m year^{-1}
Time step: 299 years
Dispersivity: 1500 m
Cation exchange capacity: 5.5 mequiv per 100 g
SO_4^{2-} reduction: 1.7 mmol l^{-1} per cell in the first four cells
Calcite equilibrium

Exchange parameters (standard coefficients in PHREEQM):
$\log K_{NaX} = 20.0$
$\log K_{KX} = 20.7$
$\log K_{CaX_2} = 40.8$
$\log K_{MgX_2} = 40.6$
$\log K_{NH_4X} = 20.6$
$\log K_{HX} = 22.5$

---

contains more Na$^+$ than could be produced by cation exchange, and it is likely that the Na$^+$ was derived from the original marine water. The highest K$^+$ concentrations, however, are associated with cation exchange that took place in the Bartonian Clay. The resulting K$^+$-rich water types have been displaced to a distance of between 13 and 5 km from the recharge area. Before K$^+$ desorption began, K$^+$ concentrations in the pore water were lower, hence K$^+$ desorption is expressed as K$^+$ peak concentrations. A similar process produced the Mg^{2+}-rich pore-water distributions; however, the Mg^{2+} desorption peak has been displaced less far from the recharge area (between 10 and 4 km) than has the K$^+$ peak. Thus the theoretical chromatographic pattern of cation exchange can still be recognized.

The deviation between model results and observations, which can be noted in Fig. 6 for the most northward part of the aquifer (the most downstream 5 km), probably results from

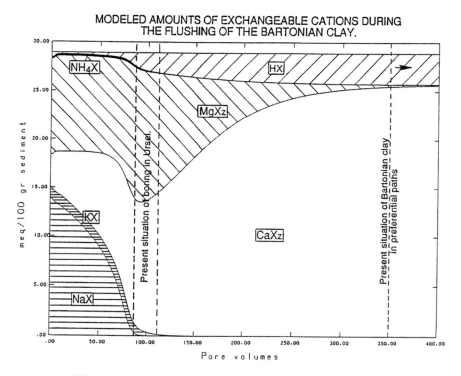
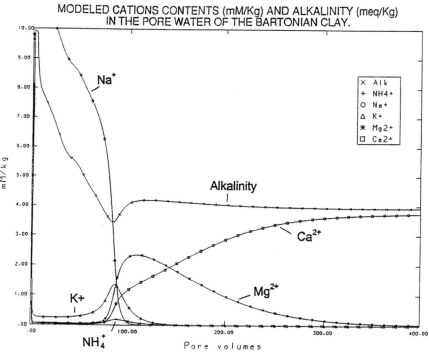

Fig. 4. Evolution of the distribution of exchangeable cations and of the composition of pore water leaking out of the Bartonian Clay during flushing.

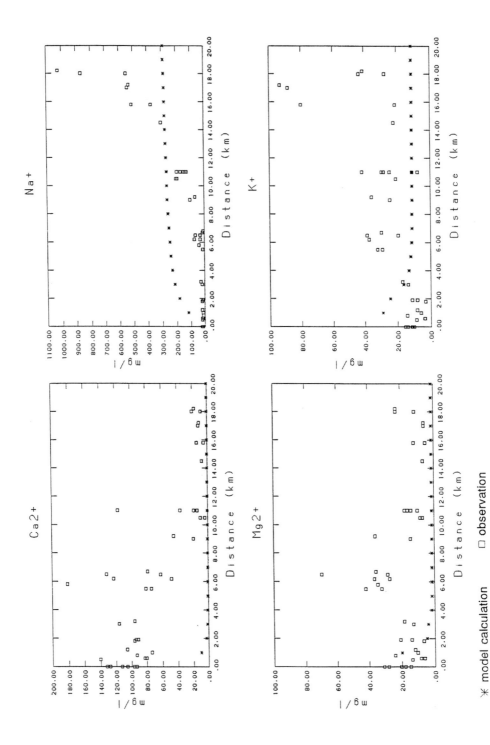

Fig. 5. Comparison of observed concentrations of main cations in Ledo-Paniselian aquifer with modelled concentrations after flushing Bartonian Clay 100 times.

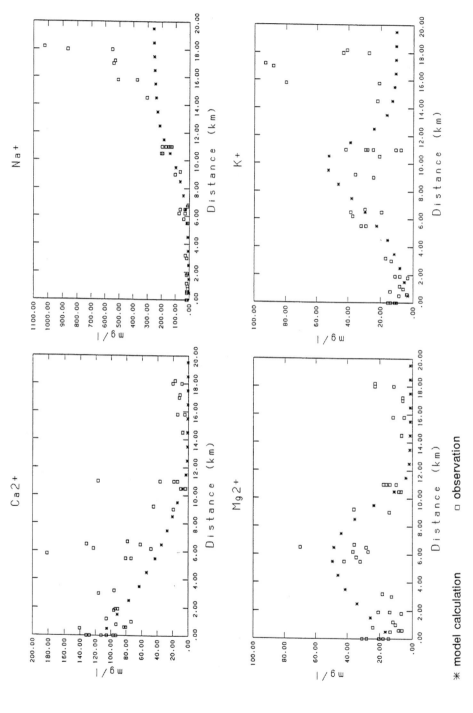

Fig. 6. Comparison of observed concentrations of main cations in Ledo-Paniselian aquifer with modelled concentrations after flushing Bartonian Clay 400 times.

the lower flow velocities in the deeper parts of the aquifer (Walraevens & Cardenal, in preparation). Lower flow velocities in this region, however, have not been accounted for in this model.

Conclusion

This study has shown that the cation exchange capacity of marine clay-rich units, through which fresh infiltration water is recharged, controls the composition of the pore water in a freshening aquifer.

A difference in hydraulic conductivity values for the Bartonian Clay of at least one order of magnitude has been obtained between laboratory measurements on local samples and a model calibration approach. Hydrogeological and hydrogeochemical evidence suggests that preferential fluid flow pathways exist in the Bartonian Clay and that the existence of these pathways could account for the discrepancy between the measured and modelled hydraulic conductivities. The apparent incompatibility between the observed adsorbed cations in the Bartonian Clay in the recharge area of the Ledo-Paniselian aquifer, and the composition of the groundwater in this aquifer, can also be attributed to the presence of preferential pathways in the clay, through which the flow is faster.

The authors are indebted to the Fund for Scientific Research, Flanders, Belgium, appointing K. Walraevens as a senior research associate, and to the University of Granada, Spain, providing a fellowship to J. Cardenal, which allowed him to stay in Belgium to carry out this research. We thank A. H. Bath for his constructive criticism, which contributed greatly to improving the manuscript, and J. Macquaker for his editorial comments and language correction.

References

APPELO, C. A. J. 1994. Cation and proton exchange, pH variations, and carbonate reactions in a freshening aquifer. *Water Resources Research*, **30**, 2793–2805.

—— & POSTMA, D. 1993. *Geochemistry, Groundwater and Pollution*. Balkema, Rotterdam.

—— & WILLEMSEN, A. 1987. Geochemical calculations and observations on salt water intrusions. I. A combined geochemical/mixing cell model. *Journal of Hydrology*, **94**, 313–330.

—— & ——, BEEKMAN, H. E. & GRIFFIOEN, J. 1990. Geochemical calculations and observations on salt water intrusions. II. Validation of a geochemical model with laboratory experiments. *Journal of Hydrology*, **120**, 225–250.

BACK, W. 1966. *Hydrochemical Facies and Groundwater Flow Patterns in Northern Part of Atlantic Coastal Plain*. US Geological Surveys Professional Paper, **498-A**.

BEEKMAN, H. E. 1991. *Ion chromatography of fresh- and seawater intrusion*. PhD dissertation, Free University of Amsterdam.

—— & APPELO, C. A. J. 1990. Ion chromatography of fresh- and salt-water displacement: laboratory experiments and multicomponent transport modelling. *Journal of Contaminant Hydrology*, **7**, 21–37.

CARDENAL, J. & WALRAEVENS, K. 1994. Chromatographic pattern in a freshening aquifer (Tertiary Ledo-Paniselian aquifer, Flanders–Belgium). *Mineralogical Magazine*, **58A**, 146–147.

CHAPELLE, F. H. & KNOBEL, L. L. 1983. Aqueous geochemistry and the exchangeable cation composition of glauconite in the Aquia aquifer, Maryland. *Ground Water*, **21**, 343–352.

FOSTER, M. D. 1950. The origin of high sodium bicarbonate waters in the Atlantic and Gulf coastal plains. *Geochimica et Cosmochimica Acta*, **1**, 33–48.

STUYFZAND, P. J. 1986. A new hydrochemical classification of watertypes: principles and application to the coastal dunes aquifer system of the Netherlands. *Proceedings, 9th Salt Water Intrusion Meeting, Delft, 1986*.

—— 1993. *Hydrochemistry and hydrology of the coastal dune area of the Western Netherlands*. PhD dissertation, Free University of Amsterdam.

TAVERNIER, R. 1954. Le Quaternaire. *In*: FOURMARIER, P. (ed.) *Prodrome d'une description géologique de la Belgique*. Société Géologique de Belgique, 555–589.

VALOCCHI, A. J., STREET, R. L. & ROBERTS, P. V. 1981. Transport of ion-exchanging solutes in groundwater: chromatographic theory and field simulation. *Water Resources Research*, **17**, 1517–1527.

WALRAEVENS, K. 1987. *Hydrogeology and hydrochemistry of the Ledo-Paniselian aquifer in East- and West-Flanders*. PhD dissertation, University of Ghent [in Dutch].

—— 1988. Application of mathematical modeling of the groundwater flow in the Ledo-Paniselian semi-confined aquifer. *In*: DE SMEDT, F. (ed.) *Computer Modeling of Groundwater Flow Problems*. V.U.B. Hydrologie, **14**, 95–113.

—— 1990. Hydrogeology and hydrochemistry of the Ledo-Paniselian semi-confined aquifer in East- and West-Flanders. *Academiae Analecta*, **52**(3), 11–66.

—— & CARDENAL, J. 1994. Aquifer recharge and exchangeable cations in a Tertiary clay layer (Bartonian clay, Flanders–Belgium). *Mineralogical Magazine*, **58A**, 955–956.

—— & LEBBE, L. 1988. Groundwater quality in the Tertiary Ledo-Paniselian aquifer in Belgium as a result of fresh-water intrusion into sediments in chemical equilibrium with the sea. *Proceedings, 10th Salt-Water Intrusion Meeting, Ghent, 1988*, 33–44.

Index

Page numbers in *italics* refer to Figures and page numbers in **bold** refer to Tables

accretionary wedges, fluid migration 36, 37–38
advection porosity 11
air, thermal conductivity **46**
air entry value 109
albite, thermal conductivity **46**
anhydrite, thermal conductivity **46**
apparent diffusion coefficient 12
aragonite, dissolution 75

Barbados accretionary complex 37, 38
Bartonian Clay *177*
   groundwater chemistry 179, *180*
      modelling 181–186
Belgium *see* Boom Clay *also* Ledo–Paniselian aquifer
bentonite
   gas permeability 113, *114*, 115–120
   shrinkage *110*
beta factor 73
Boom Clay Formation 176
   gas permeametry 113, *114*, 115–120
   porosity **17**, 18
   undrained shear deformation 64–65
boulder clay (till)
   hydraulic performance test
      method 99–101
      results 101–105
   porosity 17–18
brittle *v*. ductile behaviour 127
brittle–ductile transition 67–68
brittleness index 129, *131*
bulk density
   Conasauga Group 163
   Not Formation *143*

calcite, thermal conductivity **46**, **57**
Cambrian *see* Conasauga group
capillary sealing 125–126
carbonates
   chemical compaction 75
   diagenesis 34
   thermal conductivity **56**
cation exchange capacity 180
cementation, effect on dilatancy 129
chemistry of groundwater
   Eocene of Belgium 178–181
      PHREEQM model 181–186
chlorite, thermal conductivity **57**
clastic intrusions 37
clay
   defined 2, 24
   thermal conductivity **56**
clay minerals
   diagenesis 34
   swelling 16

claystone
   defined 2
   thermal conductivity **56**
compaction
   effect on fluid flow 75–77
   types
      chemical 75
      mechanical 73–75
compaction curves 145–146
compaction disequilibrium 145–146
Conasauga Group
   effective porosity study
      method 160–163
      results 164–165
      results discussed 165–167
   role in contaminant transport 167–170
   tectonic setting 158
connected porosity 9–10
consolidation state 98–99
consolidation testing
   experimental design 80–82
   results 82–87
contaminant transport *see* Conasauga Group

Darcy velocity 11
Darcy's Law 23, 75
deformation rate, effect on fluid flow 98
desiccation 109, 110
diagenesis
   chemical reactions 34
   effect on dilatancy 129
diffusion porosity 12–15
dilatancy onset 127–128
   effect of overconsolidation ratio 129
divided bar method for thermal conductivity 48, 50–51
dolomite, thermal conductivity **46**, **57**
ductile *v*. brittle behaviour 127
ductility estimation
   friction angle effect 129–130, *132*
   overconsolidation ratio effect 128–129
   unconfined compressive strength effect 129

effective diffusion coefficient 12
effective porosity 9, 11
   method of measurement 160–163
   results 164–165
   results discussed 165–167
embrittlement prediction
   friction angle effect 129–130, *132*
   overconsolidation ratio effect 128–129
   unconfined compressive strength effect 129
Eocene sediments of Belgium
   cation exchange capacity 180
   groundwater flow 176–178

hydrogeochemistry 178–180
  modelling 181–186
  stratigraphy 176
expansion, role in overpressure 77
extension fractures, role in leakage 126

failure envelope 68–70
failure in mudrocks 61–62
  modelling behaviour 67–70
  undrained shear deformation experiments
    Boom Clay 64–65
    Kimmeridge Clay 63–64
    London Clay 65–67
    mud volcano clays 62
    North Sea shale 65
    Todi Clay 62–63
faults
  conducting behaviour 36–38
  density mapping 133
  effect on seal leakage 130–131, *132*
  seal behaviour 36
feldspar, thermal conductivity **46**, **57**
Fick's Laws 12
Flanders *see* Ledo-Paniselian aquifer
flocculation 99
flow equations *see* fluid flow
fluid flow
  equations 108
  factors affecting 98–99
    effect of permeability anisotropy 97
  modelling with Darcy's Law 75–77
  see also hydraulic conductivity
fluid flux 11
fluid pressure studies *see* Halten Terrace
fluidization 37
fracture porosity 10
fractures
  effect on permeability 35–36
  effect on thermal conductivity 52
  role in leakage 126–127
friction angle 129–130, *132*
Fuller's Earth, thermal conductivity **50**, **51**, *57*

Garn Formation, porosity *142*
gas, thermal conductivity **46**
gas permeability
  experimental determination
    equipment 111–112
    method 113–115
    results 115–120
    results discussed 120–121
    samples 113
Gault Clay, air entry value 109
Gent Formation 176
geochemical porosity 9, 15–16
geochemistry of groundwater
  Eocene of Belgium 178–181
    PHREEQM model 181–186
gibbsite, dissolution 75
glacial clays
  hydraulic performance test
    method 99–101

results 101–105
grain density, Conasauga Group 163
grain orientation, effect on fluid flow 99
grain shape
  effect on pore size 30
  effect on thermal conductivity 52–53
grain size
  effect on pore size 30
  effect on thermal conductivity 52
groundwater flow, Eocene of Belgium 176–178

Hagen-Poiseuille equation 27
halite, thermal conductivity **46**
Halten Terrace 138
  compaction disequilibrium 145–146
  fluid pressure distribution 139–141
  fluid pressure–porosity relations 141–145
  lateral pressure transfer 147
  mass balancing in fluid flow calculation 149–150
  modelling fluid pressure buildup 146
  Rås Basin overpressure 147–149
  sedimentation rates 145
  stress history 151–152
heat flow equations 45
Heather Formation, thermal conductivity **50**
helium porosimetry 162–163
hydraulic conductivity
  effect on deformation 98
  experimental determination
    method 99–101
    results 101–105
    results discussed 105–106
  Eocene of Belgium 176–178
  London Clay *27*
hydrogeochemistry
  Eocene of Belgium 178–181
    PHREEQM model 181–186

Ile Formation, porosity *142*
illite
  precipitation 75
  role in diagenesis 34
  thermal conductivity **46**, **57**
illite–smectite, thermal conductivity **46**
immersion porosimetry 161–162
in-diffusion porosity experiments 14
internal friction 129–130, *132*

K-feldspar, thermal conductivity **46**
kaolinite
  dissolution 75
  permeability experiment
    material 80
    method 80–82
    results *82*, *83*, *88*, *89*, *90*, *91*
    results discussed 92–94
  thermal conductivity **46**, **57**
Keuper Marl, vein patterns 37
Kimmeridge Clay
  thermal conductivity **50**, **51**, *57*
  undrained shear deformation 63–64

Klakk Fault Complex *139*, 150, *151*
Knesselare Formation 176
Kortrijk Formation 176
Kozeny–Carman equation 27

landfill sites
　liner hydraulic performance 97–98
　　experimental testing
　　　method 99–101
　　　results 101–105
　　　results discussed 105–106
layering, effect on thermal conductivity 49, 52
leak-off pressure 35
leakage of top seals *see under* top seals
Lede Formation 176
Ledo–Paniselian aquifer
　cation exchange capacity 180
　groundwater flow 176–178
　hydrogeochemistry 178–180
　　modelling 181–186
　stratigraphy 176
limestone, thermal conductivity 56
London Clay
　air entry value 109
　experimental porosity measurement 14, 16, 17
　hydraulic conductivity *27*
　pore size distribution 30–32
　thermal conductivity **50**, **51**, *57*
　undrained shear deformation 65–67

Maldegem Formation 176
marine clay
　permeability experiment
　　material 80
　　method 80–82
　　results *86*, 87, 88, *92*
　　results discussed 92–94
mass balancing, fluid flow calculations 149–150
matric suction 108, 110
matrix porosity 10
mercury porosimetry 162
mica, thermal conductivity **46**
microfractures, effect on permeability 35–36
mineralogy, effect on thermal conductivity 52
moisture characteristic 109
moisture content 10
molality 9
molarity 9
montmorillonite
　compaction 73
　permeability experiment
　　material 80
　　method 80–82
　　results 82, *85*, 89
　　results discussed 92–94
mud, defined 2
mud volcano clay, undrained shear deformation 62
mudstone
　defined 2, 24
　thermal conductivity **56**

needle probe method for thermal conductivity 49, 50–51
North Sea Basin
　mudstones
　　intraformational polygonal faults 37
　　porosity data *32*
　shale, undrained shear deformation 65
　thermal conductivity **56**
　thermal history *46*, *47*
Not Formation
　bulk density *143*
　compaction curve 146–147
　formation porosity 144

Oak Ridge Reservation, Conasauga Group porosity study
　method 160–163
　results 164–165
　results discussed 165–167
　significance in contaminant transport 167–170
ODP leg 141 marine clay
　permeability experiment
　　material 80
　　method 80–82
　　results *86*, 87, 88, *92*
　　results discussed 92–94
oil, thermal conductivity **46**
Oligo-Miocene clay *see* Boom Clay
Opalinus Clay, porosity **17**, 18
orthoclase *see* K-feldspar
out-diffusion porosity experiments 14
overconsolidation ratio (OCR) 98, 128–129
overpressure
　causes 137, 140
　development in Rås Basin 148–149
　modelling buildup 146
　role in fluid flow 75–77
　role of Plio-Pleistocene loading 145–146
Oxford Clay
　porosity effects 52
　thermal conductivity **50**, **51**, *57*

Palfris Marl, porosity **17**, 18
particle orientation, effect on fluid flow 99
permeability 23–25
　anisotropy of 79–80
　　effect on fluid flow 97
　　experimental measurement 80–82
　　experimental results 82–89
　　results discussed 89–94
　diagenetic effects 34
　effect of void ratio 79
　fault effects 34–38
　pore size effects 30–34
　modelling 25–26
　　empirical 26–27
　　theoretical 27–30
　role in top seals 125
permeability *see also* gas permeability
physical porosity 9, 10–11
polygonal faults 37

pore geometry, effect on fluid flow 97
pore pressure, relation to deformation 98
pore size distribution 30–34
pore throat size 30, 163–164
porosimetry, mercury intrusion method 30
porosity
   concept 9–10
   effect of overpressure 137
   effect on thermal conductivity 52
   modelling influence of fluid pressure 146
   relation to fluid pressure 141–145
   see also effective porosity
porosity–permeability, relationship *24, 25*
pressure, effect on thermal conductivity 54
pyrite, thermal conductivity **46, 57**

quartz
   effects on thermal conductivity 52
   thermal conductivity **46, 57**

radial diffusion method 16
Rås Basin 147–148
   overpressure 148–149
relative permeability 108
rheology mapping 133

salinity, relation to geochemical porosity 16
sandstone
   compaction 74
   thermal conductivity 48, **56**
scanning electron microscopy, consolidated clays 87–89
Scottish boulder clay *see* boulder clay
seals *see* top seals
s.e.m. *see* scanning electron microscopy
shale
   defined 2
   thermal conductivity **56**
shear *see* undrained shear deformation
shear fractures
   formation 127
   role in leakage 126–127
shear zones 97
shrinkage tests 109
siderite, thermal conductivity **46**
silt, defined 2
siltstone
   defined 2
   thermal conductivity **56**
silty clay
   permeability experiment
      material 80
      method 80–82
      results 82, *84*, 87–88
      results discussed 92–94
smectite
   dissolution 75
   thermal conductivity **57**
smectite-illite, diagenesis 34
Sweden *see* Opalinus Clay *also* Palfris Marl

temperature, effect on thermal conductivity 54
Tertiary *see* Eocene *also* Oligo-Miocene
texture, effect on thermal conductivity 52–54
thermal conductivity
   experimental measurement 48–51
   mathematical expression 46
   modelling 46–48
      factors affecting 51–54
      model types 54–58
   problems in measuring 45
thermal expansion, role in overpressure 77
through-diffusion porosity experiments 13–14
Tielt Formation 176
till *see* boulder clay
Todi Clay, undrained shear deformation 62–63
top seals
   embrittlement estimation
      friction angle 129–120, *132*
      overconsolidation ratio 128–129
      unconfined compressive strength 129
   fault effects 130–131, *132*
   leak mechanisms
      effect of dilatancy 127–128
      extension fractures 126
      shear fractures 126–127
   leak risk analysis 133
   role in permeability 125
tortuosity 99
total porosity 9
transport porosity 9, 11–15
trap integrity *see also* top seals 133
Trinidad Clay **62**

unconfined compressive strength (UCS) 129
undercompaction prediction 146–147
undrained shear deformation experiments
   Boom Clay 64–65
   Kimmeridge Clay 63–64
   London Clay 65–67
   mud volcano clays 62
   North Sea shale 65
   Todi Clay 62–63

void ratio change 73, 79
void volume 9
Vøring Basin, thermal conductivity *58*

water, thermal conductivity **46**
water retention function 109
Welsh boulder clay *see* boulder clay
Whiteoak Mountain thrust sheet 158, 159
   effective porosity study
      method 160–163
      results 164–165
      results discussed 165–167
      significance in contaminant transport 167–170

Zelzate Formation 176